World Class Reliability

World Class Reliability

Using Multiple Environment Overstress Tests to Make It Happen

Keki R. Bhote

And

Adi K. Bhote

AMACOM

American Management Association

New York • Atlanta • Brussels • Chicago • Mexico City • San Francisco
Shanghai • Tokyo • Toronto • Washington, D. C.

*Special discounts on bulk quantities of AMACOM books are available to corporations, professional associations, and other organizations. For details, contact Special Sales Department, AMACOM, a division of American Management Association, 1601 Broadway, New York, NY 10019.
Tel.: 212-903-8316. Fax: 212-903-8083.
Web site: www.amacombooks.org*

Library of Congress Cataloging-in-Publication Data

Bhote, Keki R.
World class reliability : using Multiple Environment Overstress Tests to make it happen / Keki R. Bhote and Adi K. Bhote.
p. cm.
Includes bibliographical references and index.
ISBN 0-8144-0792-7
1. Reliability (Engineering) I. Bhote, Adi K. II. Title.
TA169.B48 2004
620'.00452—dc22

2003019970

Printing number

10 9 8 7 6 5 4 3 2 1

*To my beloved mother, noted author and journalist,
who hooked me on to my love of writing, and whose
"true north" life was ever my inspiration*

Contents

List of Figures

List of Tables

Preface

Despite Six Sigma, field failure rates are deplorable.

The Six Sigma initiative that we launched at Motorola eighteen years ago has won laurels all over the world. As a series of disciplines, it has far exceeded any quality movement in the last fifty years. At Motorola, we achieved an astonishing 800-to-1 quality improvement in ten years, a feat unparalleled in any company, in any country.

Yet, the customers of many companies that attempted to clone our Six Sigma process were far from satisfied. They pointedly complained that the failure rates in the field were unacceptably high. What was worse, these failure rates beyond the warranty period soared even higher; but the manufacturing companies—their suppliers—would conveniently wash their hands of all responsibility for these out-of-warranty failures.

The reason for this disconnect between low defect rates in a plant and high failure rates in a customer's field operations is the overconcentration on quality and the almost total neglect of reliability. There is a fundamental difference between quality and reliability. Quality is the goodness of the product as it leaves the factory—at zero time. Reliability has two important dimensions over quality—time and stress. Time refers to product life, where the warranty frontiers are constantly being pushed out, from one, two, and three years to seven years and even to the lifetime of the owner. Stress refers to various stresses or environmental factors, such as temperature, vibration, humidity, transients, etc. that impinge on the product and interact with one another to accelerate field failures.

The reasons for this dismal performance of reliability are well known:

- ❑ Reliability has been hijacked by mathematical jocks who are more enamored of their complex formulas than worried about unacceptably high field failure rates.
- ❑ The tools to predict, estimate, and demonstrate reliability ahead of production are embarrassingly weak.
- ❑ The American Society for Quality (ASQ) has garnered well over 150,000 members and has grown from strength to strength, while the Institute for Reliability Engineering is languishing in blackout obscurity.
- ❑ The military approach to reliability is to throw money into extensive and fruitless testing. As a result, it is estimated that it costs the Defense Department eleven times the procurement price of military hardware to maintain it—a little-known national scandal!

The Focus of This Pioneering Book

Fortunately, help is at hand. This book captures a simple, nonmathematical, low-cost but highly effective recipe to dramatically reduce field failure rates by one, even two orders of magnitude. It describes powerful tools for the twenty-first century, such as Multiple Environment Over Stress Tests (MEOST), that can predict and correct potential field failures at the design stage of the product, so companies don't have to wait helplessly to witness failures in the hands of customers. MEOST can reduce design test cost, both in labor as well as in the number of units required to demonstrate reliability. MEOST can also reduce design cycle times and beat competition to the market.

Above all, this book can show companies, mired in minuscule and fading profits, how to really make money. Most of these companies are not even aware of these profit-making techniques, much less use them, as shown later in Table 1-1.

World Class Reliability—A Chapter by Chapter Synopsis

Prologue

The book begins with an actual case study of a company, so typical of many others, that valiantly tried but failed to achieve its targeted reliability—at a cost of extreme customer disillusionment and grievous hits to the pocketbook. This story sets the stage for what must be embraced before achieving a true reliability breakthrough and soaring profits.

Part 1: The Challenge for Industry—Regaining Lost Profits

Chapter 1 attempts to transform a corporate paradise lost to a paradise regained. With a sagging economy, intense worldwide competition, and a looming specter of deflation, companies today are thrashing around in pathetic and futile cost-reduction measures. Instead, companies desperately need to understand at least ten proven techniques—ten gold nuggets—that can dramatically raise their profits to unbelievably high levels.

It is not the purpose of this book to detail each gold nugget. A summary must suffice as a guide. However, one of these techniques—reliability breakthrough—is the dominant theme of this entire book.

Chapter 2 details the compelling need for reliability, the reach-out objective of reliability, and the overwhelming benefits of reliability.

Part II: What Not to Do—Current but Ineffective Reliability Methodologies

Chapters 3–7 trace the sorry state of the reliability disciplines in vogue. Chapter 3 shows the weakness of reliability mathematics that are complex and unproductive. Chapter 4 exposes reliability prediction techniques that miss the mark. Chapter 5 shows how reliability estimators are lousy forecasters. Chapter 6 critiques reliability demonstration that wastes money and time.

Finally, Chapter 7 develops the transition to two newer techniques: HALT (Highly Accelerated Life Tests) and HASS (Highly Accelerated Stress Screening). They have advanced the reliability state-of-the-art, but have weaknesses that limit their effectiveness.

Part III: The Climb to MEOST—The Mount Everest of Reliability

Chapter 8 is base camp 1, where we begin our preparation for the Multiple Environment Over Stress Test (MEOST). The first effort is setting up a much-needed infrastructure for design reliability. That infrastructure includes organizing the design function; implementing important management guidelines and audit; capturing customer "wow"; establishing realistic specifications and tolerances; and releasing the bottled-up creative genius of the engineer.

Chapter 9 is base camp 2, where we inspect the essential prerequisites for MEOST. The details include thermal scanning, modular designs, and design for manufacturability. Other supporting but important disciplines are highlighted within the design function. These disciplines include derating; built-in diagnostics; product liability analysis; poka-yoke; design for robustness, resale; maintainability and availability; variation reduction (high C_{pk}s); and design robustness.

Similar disciplines in manufacturing and in the field include destructive process analysis (DPA); failure analysis; field escape control; defect minimization; zero time failures; and field data retrieval.

Chapter 10 is the final climb to the Mount Everest of reliability. It defines the salient principles of MEOST and describes its history, applications in industry, and its successes. After covering the preambles and preparations of MEOST, we reach, finally, the pièce de résistance—the eight stages of MEOST.

From this vantage point, Chapter 11 surveys the versatility of MEOST. Its techniques are applied in several industrial situations, from preventing lawsuits to maximizing resale value, validating design improvements, gauging out-of-warranty performance, and extending MEOST to key suppliers.

Finally, the Appendix covers the characteristics and requirements of the chambers and equipment needed for a successful MEOST launch.

Throughout the text, there are case studies to illustrate the limitation of traditional reliability techniques and the power of the practical, surefire techniques such as MEOST.

World Class Reliability

Prologue

The Perfect Circle Corporation: A Case Study Background*

Perfect Circle is a large appliance manufacturer in the United States with a few subsidiaries in Western Europe and Asia. It has a dozen manufacturing facilities in the States dedicated to specific appliances. Located, for the most part, in smaller cities and towns, the company has a stable, loyal workforce. It is unionized, but the union's leadership works in relative harmony with the plant management.

The Six Sigma Initiative

Several years ago, the company embarked on a comprehensive quality improvement initiative. It hired a Six Sigma consulting company to facilitate the program.

*The name is fictitious, but the case study represents a real company, with real reliability problems.

The consulting company launched a torturously long training curriculum, extending to six whole months. The first six weeks were spent in management orientation and championship training. These management graduates were to guide master black belts, who would act as internal consultants and coaches to train their own corps of black belts in solving chronic quality problems. The black belts had four rounds of one-week instruction and three weeks of practice for each round.

- The first round was focused solely on measurement accuracy.
- The second round of instruction was on finding causes for problems using ineffective cause-and-effect diagrams.
- The third round was the Define, Measure, Analyze, Improve, and Control routine (DMAIC), which provided only a skeleton framework for problem solving. The principal technique used was a complicated, cumbersome, and statistically weak Classical/Taguchi Design of Experiments.
- The fourth round attempted to control the process by documenting procedures and using the obsolete tools of control charts.

There was no emphasis on:

- The customer and how to retain his loyalty
- Reliability and methods for reducing failure rates in the hands of customers
- Attributes that spell the difference between leadership and management

- Organization infrastructure to dismantle Taylorism
- How to release the creative potential of employees and move to meaningful empowerment
- How to secure real partnerships with key suppliers/ distributors and improve quality, cost, and cycle time
- Design for robustness, for manufacturability, for minimum variation, and for built-in diagnostics
- Business processes tackled with out-of-the-box thinking and innovation

While modest improvements in quality were registered, the cost of the consultant training—$50,000 for each black belt student, which totaled $5 million for the 100 black belts—so disillusioned the company that it sent the consulting company packing and settled on a homegrown approach. All the while, Perfect Circle's customers, who expected better field results from the consultant Six Sigma program, were equally disillusioned. Something had to be done—and quickly—to restore the company's reputation.

Year 0—The 10:1 Goal of Field Reliability Improvement

At its annual management retreat, company management unfurled—with a great deal of fanfare—a reliability campaign for a ten-to-one reduction in field failure rates in ten years, using its current failure rate of 13 percent per year as a baseline. (We at Motorola would have committed hara-kiri with such astronomically high failure rates. Today, our failure rates are less than 0.01 percent, or 100 parts per million per year.)

Management set the goal, but left execution to lower levels. Furious activity commenced. Tools such as Pareto charts, Ishikawa diagrams, brainstorming, and process mapping were employed. What follows is the company's reliability timeline.

> Perfect Circle's approach to reliability is reminiscent of a story of a convention of cockroaches looking for ways to prevent being crushed to death by Homo sapiens. The convention chairman announced the solution of the executive committee. "We resolve to transform ourselves from cockroaches to grasshoppers. With that greater mobility we can survive." The resolution was thunderously approved. Then one lone cockroach stood up and asked: "Exactly how do you propose to make the transformation from cockroaches to grasshoppers?" Whereupon the chairman proudly announced: "Ours is an executive committee. We'll leave execution to the lower levels!"

Year ½—Improvement on Track

At the end of six months, management proudly announced an overall failure rate projected at 11.8 percent per year, an encouraging reduction of 9 percent from the baseline 13 percent level. This was hailed as terrific progress. There were estimates that the rate of improvement was on target to meet the goal of a 1.3 percent failure rate in ten years. Management congratulated the team and urged it to continue the momentum.

Year 1—An Unexplainable Failure Increase

When the tally was counted at the end of the first year's reliability drive, the failure rate actually increased to 12.1 percent. The setback was explained away as "noise" in the field data-gathering system. The 7 percent improvement was still considered meaningful.

Year 1½—Another Small Step for Reliability

One-and-a-half years later, the field failure rate was reported at 11.5 percent—a 12.5 percent improvement over the baseline figure of 13 percent. The company believed the numbers were moving in the right direction, though perhaps a bit slower than the targeted improvement.

Year 2—The Reliability Needle Stuck at 11.7 Percent

Storm clouds were gathering. The failure rate needle had hardly moved in a year—from around 11.8 percent to 11.7 percent. It is well known in industry that small fluctuations in these types of numbers are statistically insignificant to register improvement or deterioration. But the finger-pointing started. "Our dealers are ripping us off" and "Our installers are untrained" were some of the excuses.

Year 2½—Field Failure Rate at 10.9 Percent

A sigh of relief. "Maybe we're making headway after all" was the self-congratulating response.

Year 3—A Yo-Yo Failure Rate at 11.8 Percent

Three years after the reliability kickoff, the failure rate that should have dropped to 9 percent—a 30 percent improvement—was barely registering a 9 percent improvement. At this point, the fur really started to fly.

> "Our engineering changes have not had enough field exposure."
>
> "Our customers are too finicky."

"Our customers do not understand the complexities of our controls."

"Our customers misuse our appliances."

And the mother of all excuses:

"Twenty-five percent to 30 percent of our returned units are NADs—No Apparent Defects. If these are subtracted from the failure rates reported, we would have been on target!"

Year 4—Recalls and a Lawsuit

The coup de grâce for the reliability improvement program came when customer complaints reached such a crescendo that the company was forced to recall two of its products at a cost far exceeding warranty payments. One of the customers filed a lawsuit for personal damage, and its ambulance-chasing lawyers had a field day in explaining the case in a class-action litigation. A year after the lawsuit was filed, the outcome has not been settled in court.

Moral of the Case Study—What Not to Do

This case study offers many lessons on what not to do when reliability improvement is your goal:

- ❑ Do not hire hyped Six Sigma consultants who, with little depth of knowledge of the real Six Sigma, overpromise and under-deliver.
- ❑ Do not create an elitist black belt cadre. Why not convert a whole factory into a black belt army with simple

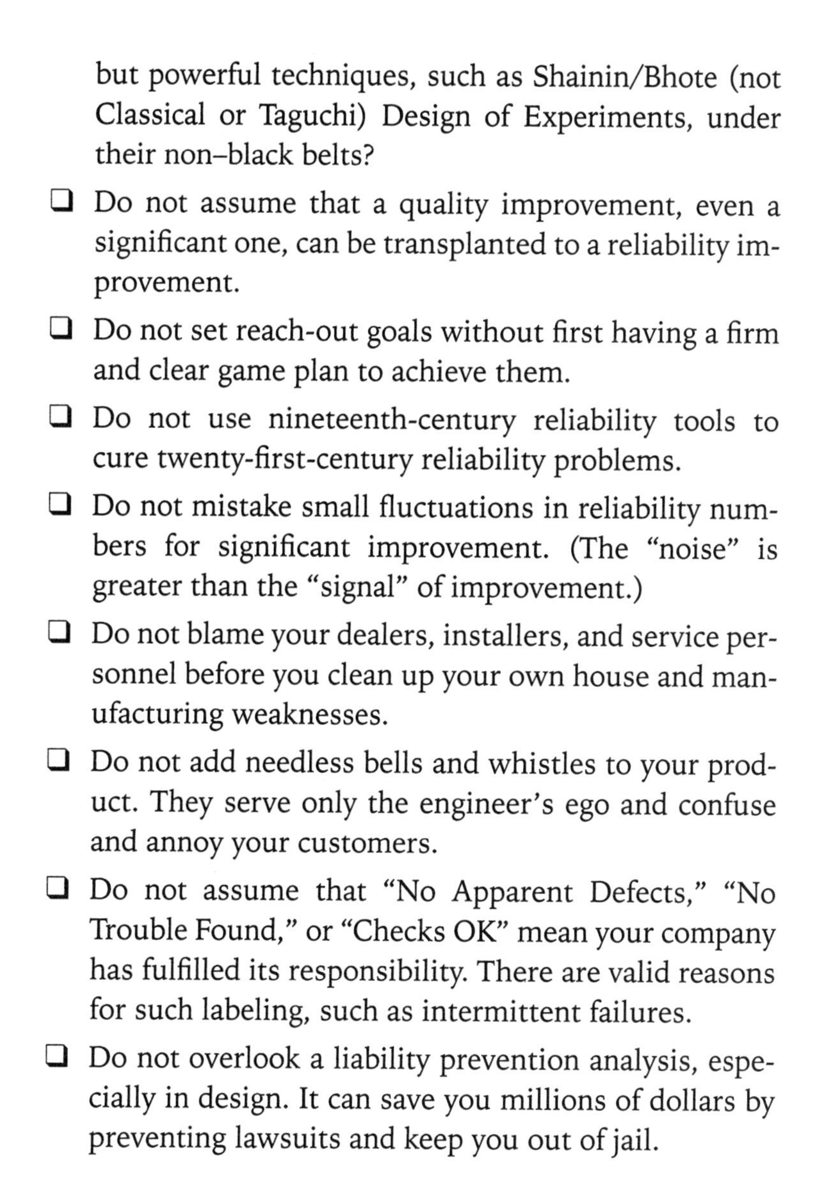

but powerful techniques, such as Shainin/Bhote (not Classical or Taguchi) Design of Experiments, under their non–black belts?

- ❑ Do not assume that a quality improvement, even a significant one, can be transplanted to a reliability improvement.
- ❑ Do not set reach-out goals without first having a firm and clear game plan to achieve them.
- ❑ Do not use nineteenth-century reliability tools to cure twenty-first-century reliability problems.
- ❑ Do not mistake small fluctuations in reliability numbers for significant improvement. (The "noise" is greater than the "signal" of improvement.)
- ❑ Do not blame your dealers, installers, and service personnel before you clean up your own house and manufacturing weaknesses.
- ❑ Do not add needless bells and whistles to your product. They serve only the engineer's ego and confuse and annoy your customers.
- ❑ Do not assume that "No Apparent Defects," "No Trouble Found," or "Checks OK" mean your company has fulfilled its responsibility. There are valid reasons for such labeling, such as intermittent failures.
- ❑ Do not overlook a liability prevention analysis, especially in design. It can save you millions of dollars by preventing lawsuits and keep you out of jail.

Finally, remember: The cost of one lost customer over his lifetime can be 100 to 1,000 times the cost of a warranty claim.

PART I

The Challenge for Industry: Regaining Lost Profits

CHAPTER 1

Corporate Paradise Lost and Regained—The Ten Gold Nuggets for a Dramatic Profit Increase

"Most companies are so mesmerized by the lure of high profits that they achieve neither customer loyalty nor breakthrough profits nor that sacred cow—shareholder value."

—KEKI R. BHOTE

The Alphabet Economy

There seems to be an "alphabet soup" syndrome to depict the projected shape of the U.S. economy:

V Cloud-nine economists who predict a sharp and steep recovery
U Economists with rose-colored glasses, who predict a more graduated but eventually robust turnaround

L Realistic economists who predict a flat but high jobless-ness rate, after the recession, that may prolong for a few years
W Pessimistic economists who predict an eventual recovery, preceded by a double-dip recession
M Gloom-and-doom economists who predict a double-dip recession—period

Whatever the eventual outcome, there is no question that the present economy is sluggish. The fingerprints are all over the landscape:

- ❑ Persistent unemployment
- ❑ Rising bankruptcies
- ❑ Falling consumer confidence
- ❑ Swollen consumer debt
- ❑ A stock market that rises only to fall the next day

Business Management—The Bête Noire, the Culprit

There are several reasons for this malaise. That robust engine of economic development—the American consumer—is pulling in his buying spree horns. The impact of 9/11 is still reverberating; the threat of al-Qaeda terrorism seems to hover over the land. The federal government's fiscal and monetary initiatives are in limbo, not advancing jobs. But the main cause is the sorry state of business and industry. And within that framework, it is the *management* of companies that is the bête noire (black sheep), the root cause, or—as we say in the discipline of the Design of Experiments—the Red X.

One perspective on management is the motley crowd of

CEOs, accountants, and external auditors who have been found guilty—at least in the court of public opinion and the stock market—for their unseemly greed, their deplorable ethics, and their warped values.

A second perspective is management's knee-jerk, blind, and counterproductive response to a downturn in their companies' fortunes. Those responses have included:

- *Layoffs*. This is the most common approach, even though many studies have revealed that the bottom line has not improved with such disruptions to the organization's culture. To add insult to injury, the job cuts at the lower levels are accompanied by huge bonuses for top management.
- *Freeze on All Travel*. Business trips to visit partnership suppliers and customers are outlawed, at a time when they may be needed more than ever.
- *Payroll Cuts Across the Board*. These cuts are so indiscriminate that they may be cutting into the corporate bone, not just the fat. It's a recipe to lose the most valuable employees, who should be retained and nurtured.
- *Move to Offshore Manufacturing*. It is well known that going abroad makes sense only if direct labor costs are greater than 8 percent of sales. The great majority of manufacturing companies, especially in assembly-type industries, have direct labor costs below 5 percent of sales. Furthermore, in the unseemly rush to offshore countries such as China, the hidden costs of poor quality, freight, customs duty, and many other factors can negate the so-called savings in direct labor.
- *Automation*. Some companies turn to automation, instead of offshore manufacturing, as a means of reducing direct labor costs. It is the wrong direction to take. It merely transfers cost from direct labor to maintenance and adds inflexibility,

to boot.

- *Mergers.* There is a mad scramble for companies to merge with others for complementary advantage. Yet the road to mergers is paved with disappointment and disillusionment. It is difficult enough for one company to sustain a culture of enthusiasm and productivity. It is almost impossible to merge two such cultures without serious injury to the merged people—their hopes and their future.

Ninety Percent of Management Does Not Know How to Make *Real* Profits!

The final perspective is the unkindest cut of all. Top management, while worshiping the sacred cow of profit and mouthing the mantra of shareholder value, does not know how to maximize, or even optimize, either profits or shareholder value.

The Ten Gold Nuggets to Make Profits Soar

There are ten disciplines/techniques—gold nuggets, we call them—that we have distilled from a larger list of 200 disciplines that can skyrocket a company's profits. (For a complete treatment of these disciplines, see Keki R. Bhote's *The Power of Ultimate Six Sigma,* New York: AMACOM Books, 2003.) The tragedy is that most senior managers are not even aware of them, much less using them.

Table 1-1 is a list of these ten disciplines. The table notes the range of profit improvement each discipline can create, along with the tiny percentage of companies that are aware of the disciplines and the even tinier percentages that implement them. Each of these disciplines is then briefly explained (more details would require a whole book for each), except for discipline 4—reduction in field failure rates—which forms the subject of the remaining chapters in this book.

Table 1-1. Disciplines to increase corporate profitability.

Technique/Discipline	*Range of Profit Improvement*	*% of Companies Aware of the Discipline*	*% of Companies Implementing the Discipline*
1. 5% Increase in customer loyalty/retention	35% to 120%	< 5%	< 0.5%
2. 20% Reduction in customer base	25% to 40%	< 1%	< 0.1%
3. 50% Reduction in cost of poor quality	100% to 200%	< 20%	< 5%
4. 50% Reduction in field failure rates	50% to 150%	< 1%	< 0.01%
5. 50% Improvement in overall equipment effectiveness (OEE)	20% to 75%	< 10%	< 1.5%
6. 10% Reduction in bill of materials	80% to 120%	< 50%	< 10%
7. 3:1 Increase in inventory turns	30% to 75% (ROI)	< 50%	< 5%
8. 2:1 Reduction in design cycle time	1 to 2 years improvement in cash flow	< 10%	< 2%
9. 2:1 Reduction in business process cycle time	20% to 75%	< 15%	< 2%
10. 25% Cost reduction through value engineering	20% to 40%	< 10%	< 2%
Overall Potential for Improvement	2:1 to 4:1	Average: < 10%	Average: < 2%

Granted, there is some degree of overlap and interaction between these disciplines. But simple addition reveals that a 2:1 improvement in profits would be a 100 percent certainty, given that the average profit after tax today for a company is less than 4 percent. The potential for an even higher

3:1 profit improvement has a high probability. Finally, reaching out for a meteoric 4:1 profit improvement is within the realm of possibility.

Discipline 1: Customer Loyalty/Retention[1]

Many companies have jumped on the customer satisfaction bandwagon. Yet, most of them do not realize that the customer train has left the station without them. What are the facts on customer satisfaction?

- ❑ According to the Juran Institute, fewer than 2 percent of companies are able to measure bottom-line improvements from increased customer satisfaction levels. There is little correlation between customer satisfaction and profit.
- ❑ Between 15 percent to 40 percent of customers, who say they are satisfied, will defect from a company each year.
- ❑ In the auto industry, the average repurchase rate of satisfied customers of the same carmaker is less than 30 percent.

What are the facts, by contrast, on customer loyalty?

- ❑ There is close correlation between customer loyalty and profit.
- ❑ A 5 percent reduction in customer defection generates a 35 percent to 120 percent profit increase.
- ❑ One lifetime customer is worth more than $850,000 to a car company.
- ❑ Loyal customers provide higher profits, more repeat business, higher market share, and more referrals than do "just satisfied" customers.

There are five key factors that are essential for a company to maximize customer loyalty, customer retention, and customer longevity.

Success Factor 1: Measuring the Cost of Customer Defection

Most companies do not measure the defection rate of their core customers. Worse, they do not even know how to measure, let alone analyze, this metric. This is one of the worst profit leaks in a company. Yet, it is actually easier to measure customer defection than customer satisfaction.

Success Factor 2: Capturing Customer "WOW"

There are many elements that add up to customer "wow"—quality, reliability, price, delivery, safety, technical "bells and whistles," to name a few. Two elements are especially important: 1) any element missing from a company's product that is important to the customer; and 2) any unexpected features that surprise and delight customers.

Success Factor 3: Taking the Customer's Skin Temperature Every Day!

This activity entails:

- ❑ Forming a win-win partnership with core customers
- ❑ Taking the pulse of each core customer—his or her needs, concerns, irritations—at all times, and especially after the sale is consummated
- ❑ Establishing close personal relations, based on mutual trust

- Arranging visits by top management to get unfiltered customer feedback

Success Factor 4: Establishing a Company Infrastructure for Customer Loyalty

Key elements of infrastructure include:

- A top management steering committee to ensure long-term customer retention
- A chief customer officer (CCO) acting as customer czar
- Meaningful metrics to track customer retention
- A high-octane SWAT team to analyze and reduce customer complaints and defections

Discipline 2: Customer Base Reduction—Customer Differentiation[2]

While companies have heeded Peter F. Drucker's advice to make a "concentration decision" and reduce their supplier base, they continue to sweep the floor and haul in as many customers as possible. The truth of the matter is that *not all customers are worth keeping.* Differentiation means separating your customers into categories:

- *Platinum and Gold Customers* constitute only 25 percent of the total customer base by volume, but produce almost 70 percent of the company's profits. They are the crown jewels, the most loyal of customers and the most difficult for competition to dislodge.
- *Silver Customers* account for 35 percent by volume and 30 percent of the profits. They are worth cultivating,

but not as assiduously as the platinum and gold customers.

- *Bronze and Tin Customers* occupy 40 percent by volume, but actually drain profits by 20–30 percent. If they cannot be reformed, make a present of them to your competition.

Increasing sales willy-nilly is not the answer. Profit is the name of the game. Differentiation allows you to increase your profits by 25 percent to 40 percent.

Discipline 3: Reduction in the Cost of Poor Quality[3]

Cost of poor quality (COPQ) can be defined simply as external failure costs (warranty); internal failure costs (scrap, analyzing, and rework); appraisal costs (all inspection and test); and cost of inventory. Added up, these COPQ costs range from at least 10 percent to 20 percent of a company's sales dollar. This is a total loss! What's more, over 80 percent of companies are unaware of this loss. Another 15 percent do not measure it. Another 4 percent do not analyze it or reduce it. Yet, what a fantastic profit-making tool COPQ can be for a company. Even a modest 50 percent reduction in COPQ can take typical profit on sales from 4 percent to between 9–14 percent—which is a 2:1 to 3.5:1 profit increase.

The specific tools, besides overall quality management practices, for reducing COPQ are:

- Design of Experiments—Shainin/Bhote[4] (not Classical or Taguchi)—for chronic quality problem solving and problem prevention
- Multiple Environment Over Stress Tests (MEOST) for reliability breakthrough

- ❑ Voice of the Customer, for mass customization/quality function deployment
- ❑ Realistic specifications and tolerances derived from the customer
- ❑ Total Productive Maintenance[5,6] (see Discipline 5)
- ❑ Poka-Yoke,[7,8] for preventing operator-controllable errors

Discipline 4: Reduction in Field Failure Rates

This discipline is the focus of this book and will be detailed in the subsequent chapters.

Discipline 5: Improvement in Overall Equipment Effectiveness (OEE)

Total productive maintenance (TPM) was introduced in Japan nearly twenty years ago to improve process and machine productivity. Its metric, OEE, is calculated using the formula:

$$\text{OEE Percent} = \text{Product Yield (as a percentage)} \times \text{Equipment Uptime (as a percentage)} \times \text{Machine Efficiency (as a percentage)}$$

where machine efficiency can be simply defined as:

$$\text{Theoretical runtime}/(\text{Theoretical runtime} + \text{Setup time})$$

These three percentages multiplied together should produce an OEE of at least 85 percent to be world-class. In most companies, the metric is not even known. In a few companies where it is measured, OEE is 20–50 percent. At the 85 percent or higher level, a manufacturing operation can expect a minimum savings of 4–7 percent of sales (i.e., almost a doubling of profits).

The specific tools to achieve these high OEE levels are:

OEE Metric	Tool
Product Yield	Design of Experiments (DOE), process certification, and positrol
Equipment Uptime	DOE, preventive maintenance (not breakdown maintenance), and process certification
Machine Efficiency	Setup time reduction, DOE

Discipline 6: Reduction in Bill of Materials Each Year[9]

The truly important area for cost reduction isn't in production (where direct labor accounts for less than 5 percent of sales) but in materials from suppliers, which constitute 50 percent to 60 percent of sales. That's a leverage of ten times the corresponding impact of cost reduction in production alone.

There are several success factors leading to a continuous 5 percent to 15 percent reduction in purchased material costs each year. They include:

1. Establishing a win-win partnership with key suppliers, instead of engaging in a confrontational win-lose contest
2. Negotiating supplier prices as a *ceiling* and supplier profits as a *floor*
3. Outsourcing all products/functions that are strategically unimportant and where the company does not have great competency

4. Having a top management steering committee to provide focus and guidance (a great necessity), and using commodity teams as the workhorse of supply chain management
5. Requiring early supplier involvement at the prototype stage of design
6. Cost targeting—where the company, not the supplier, calculates a target price
7. Establishing financial incentives/penalties, especially for meeting (or failing to meet) agreed-upon reliability targets
8. Focusing on supplier development and active, concrete help in achieving quality, cost, and cycle time improvement

Discipline 7: Increase in Inventory Turns[10]

Most companies have inventory turns (i.e., sales dollars/year divided by average inventory) in the single-digit range or in the teens, at best. World-class companies achieve 60 to 120 turns (i.e., a total cycle time from raw materials to customer shipments of less than one week or even three days).

There are three major areas for inventory turn increases: 1) work-in-process, 2) raw materials from suppliers, and 3) finished goods to customers.

Work-in-Process (WIP)

- ❑ Pull (kanban) vs. push (MRPI) systems
- ❑ Focused factory
- ❑ Product vs. process flow
- ❑ Setup/changeover time reduction

- ❑ Preventive vs. breakdown maintenance
- ❑ Small lots and linear output
- ❑ Total productive maintenance (TPM)

Raw Materials from Suppliers

- ❑ All of the WIP success factors applied to key suppliers
- ❑ A, B, C stratification of parts
- ❑ Partial authorization of subsupplier bottleneck parts inventory
- ❑ Blanket orders with volume-variable pricing

Finished Goods to Customers

- ❑ Negotiations to request customer orders in small, frequent amounts
- ❑ Partial authorization of long-lead-times inventory
- ❑ Blanket orders with volume-variable pricing
- ❑ Elimination of distributor/dealer stocking; direct shipment to point of sale

Discipline 8: Reduction in Design Cycle Time[11]

This is one of the most fruitful areas of profit improvement because:

- ❑ Anywhere from 70–80 percent of quality problems originate in design.
- ❑ Eighty percent of high costs are traceable to design.
- ❑ One to two years of outward cash flow is expended in design before a company can recover it through accounts receivable.

The success factors in design are explored in depth in Chapter 8 on design reliability infrastructure.

Discipline 9: Reduction in Business Process Cycle Time[12]

Cycle time can be defined in quasi-mathematical terms as:

$$\text{Cycle time} = \int \text{Q, C, D, E}$$

where cycle time is the integral of quality (Q), cost (C), delivery (D), and effectiveness (E). Any improvement of these last four factors results in a reduction of cycle time, which therefore can be used as a single metric to monitor improvements in major business processes.

Success factors include:

1. *Organization* (steering committee, process owner, cross-functional teams)
2. *Techniques* (Next Operation as Customer, or NOAC)

- ❑ Internal customer as evaluator and scorekeeper of internal supplier
- ❑ Ten-step NOAC process (flowcharting)
- ❑ Out-of-box thinking (Total Value Engineering, DOE, process redesign)

Discipline 10: Overall Cost Reduction[1]

There are general approaches that provide the infrastructure for meaningful cost reduction, yet they are little known or seldom employed in most companies. There are specific tools that are even less known or used.

Part A. General Approaches

1. *Leadership vs. Management.*[1] The latter is a dirty word; the former—an ennobling characteristic.
2. *Organization.* To effect a culture change, people's beliefs and values must be changed. To change people's beliefs and values, the management system and organization must be changed:

   - ❑ From a tall organizational pyramid to a flat structure
   - ❑ From departmental "silos" to cross-functional teams
   - ❑ From the chains of Taylorism to freeing the human spirit
   - ❑ From mind-numbing rules and regulations that do not concern the customer to a focus on customer loyalty
   - ❑ From hiring/selecting narrow skills to broad-based education
   - ❑ From performance appraisals that are boss/supervisor evaluations to customer assessment and 360-degree appraisals
   - ❑ From pro forma merit raises (minuscule raises with little differentiation) to incentives/bonuses for performance

3. *The Mini-Company.* Run a piece of the company as an autonomous enterprise.

Part B. Specific Disciplines

- ❑ Total Value Engineering (used to maximize customer "wow" at minimum cost)
- ❑ Job redesign (used to strengthen line functions and attenuate staff functions)
- ❑ SBU, product, model, and part number reductions
- ❑ Group technology (used to reduce part number proliferation)
- ❑ Early supplier involvement in design
- ❑ Financial incentives/penalties (especially for reliability)

Conclusion

This high-octane mix of disciplines/techniques can, quite amazingly, transform a corporation from a money loser or an also-ran to a company with world-class profitability. By following these disciplines, your company can easily outpace many of the current companies making up Fortune's list of the 100 most successful companies in the United States.

CHAPTER 2

A Breakthrough in Reliability—The Need, Objectives, and Benefits

"A cartoon depicting a New Yorker's perception of U.S. geography shows Manhattan occupying half the United States, with a tiny city in the west called Hollywood, and nothing but prairie land in between.

A similar outlook by corporate America on quality would reveal a large subcontinent called 'quality,' with an obscure town called 'reliability' elsewhere."

—KEKI R. BHOTE

The Compelling Need for Reliability

1. The Perception of Reliability Subsumed in an All-Powerful Definition of Quality

The word *quality* has been a household term—a synonym for everything that is "good." It is pervasive, growing daily in stature and importance. Reliability, on the other hand, is little known, ill-defined, and at best, relegated to be a mere handmaiden of quality. Consider these facts:

- In the well-known quality standards and systems, such as ISO-9000, the Malcolm Baldrige National Quality Award, and the Deming Prize, reliability does not even rate a mention.
- The American Society for Quality (ASQ) commands a membership of over 150,000, while the Institute of Reliability Engineers withers on the vine.

Yet, if the main purpose of a business is customers (and not quality, as many of its practitioners assert), then reliability plays a much larger role in customer expectations.

The confusion in terminology between "quality" and "reliability" goes back to the very definition of these two concepts.

- Chronologically, quality is associated with a product as it exits the plant door—at time (t) = 0.
- Reliability has two important dimensions beyond quality—time and stress. A product has to live for one, two, five, and even twenty years in the hands of the customer. The product must also function despite life-threatening stresses applied to it, such as temperature, vibration, shock, voltage transients, and several other environments.

2. The Monumental Cost of Unreliability

Many companies take a cavalier attitude in dismissing their costs associated with reliability. They express their satisfaction if warranty costs are within one percent to 2 percent of their sales dollars. They even justify such costs by baking them into their budgets.

Tragically, most companies do not realize that warranty costs are only the tip of the iceberg (see Figure 2-1). Added to warranty costs are the costs of complaints, repair service,

Figure 2-1. Warranty cost—just the tip of the iceberg (a ratio of 1:10 up to 1:100).

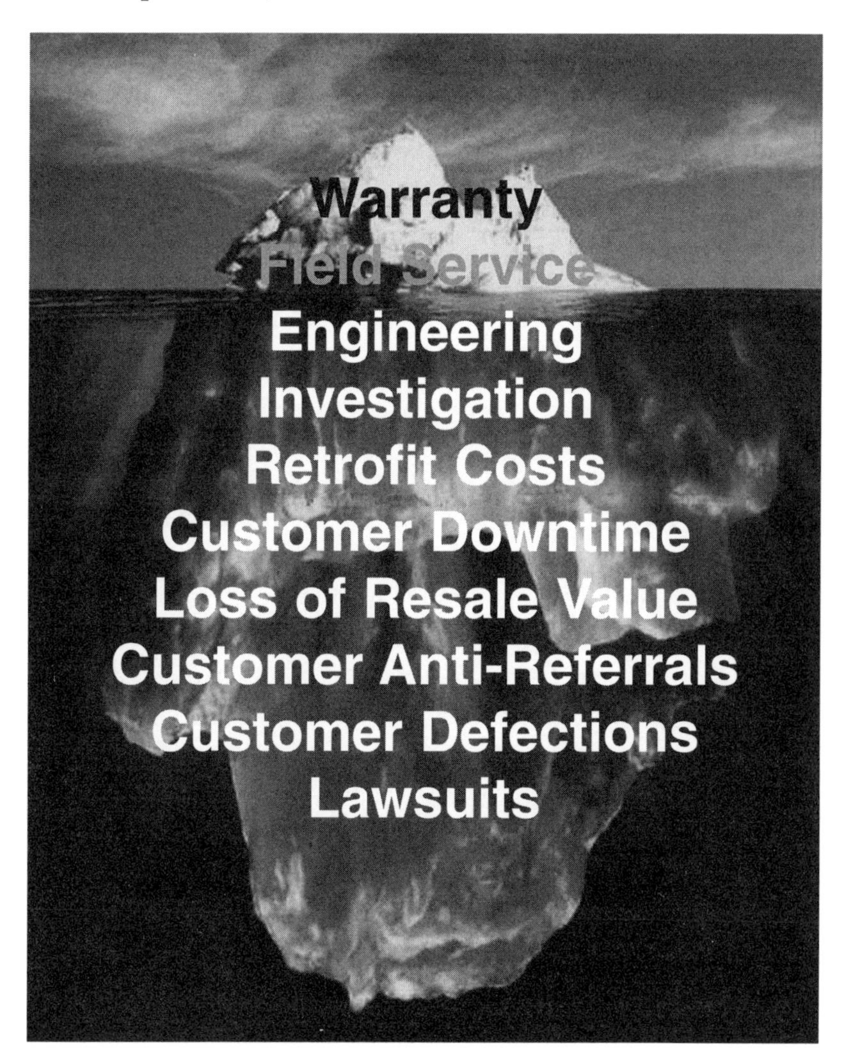

retrofits, recalls, and lawsuits. The "coup de grace" is delivered when the customer walks away from the company and into the arms of its competitor—forever. Here's an example.

Warranty Costs for the Big Three Automakers/Year

- ❑ Per-car cost: $462
- ❑ Total cost: over $6 billion
- ❑ Resale value loss caused by poor reliability and image: U.S. vs. Japan: $4,800 per car
- ❑ Consumer compensation for low resale value, through incentives that average more than $2,600 per car: a loss of more than $30 billion a year

It is estimated that the loss of one customer for a car company can be $950,000. The rationale for this astounding figure is as follows:

> Assume that the customer is so thoroughly disgusted with the automaker that he vows never to buy a car from the company ever again. Assume also that he is a young man of twenty-five years and is likely to buy ten new cars at an average price of $30,000 in his adult life. Let us further assume that he complains bitterly to twenty of his friends and relatives about his bad experience and that three of them follow his example of not buying from the company. The cost to the automaker is not the sales loss of one car, but ten cars times $30,000 times three, which equals $900,000. Furthermore, the loss of financing, service, and parts could add at least another $50,000 for the lifetime of the customer.

3. A Rising Chorus of Customer Complaints

Customers have responded to this lack of reliability in growing numbers, as evidenced by these statistics in several product categories:

A. General

- General products returned by consumers to retailers as unreliable have steadily gone up from 59 percent twenty years ago to over 70 percent today, while direct complaints to manufacturers have increased from 19 percent to over 30 percent.[13]

B. Automotive

- More cars are recalled in one year than are built.
- The U.S. Office of Consumer Affairs receives more complaints on automobiles than any other product.

C. Military

- Electronic equipment used by the Navy operates only 30 percent of the time.
- Two-thirds to three-fourths of electronic equipment is either out of commission or in repair, according to the Army.
- Maintenance costs of electronic equipment is eleven times the contract (procurement) price in five years, according to the Air Force.
- Maintenance costs of electronic equipment used by the Navy is twice the contract price in the first year.

4. The Growth of Consumerism

Poor reliability has created a backlash among consumers, leading to Ralph Naderism, consumer protection societies, and the creation of an Office of Consumer Affairs Ombudsman.

5. "Big Brother" Legislation

Increasingly, federal and state laws are closing in on poor reliability with:

- ❑ Simplified warranties
- ❑ Return obligations (i.e., lemon laws)
- ❑ Consumer protection
- ❑ Product recalls

6. Product Liability and "Ambulance Chasing"

More and more, product liability lawsuits are being settled out of court, with companies forced into "nuisance payments" to avoid the enormous costs of such suits dragging on in the courts for years on end. The whole emphasis has shifted from "caveat emptor" to "caveat vendor"—from let the buyer beware to let the manufacturer beware. Frivolous punitive damages are imposed by juries guided not by the justice of the case, but by the ability of the big company to pay "the little old lady in tennis shoes."

7. Overseas Competition

The rising economic power of the European Union and the strong quality/reliability thrust of the Japanese companies (who are getting stronger while the Japanese government gets weaker) are signs of impending danger that U.S. companies are not heeding. Witness the Japanese increasing their share of the car market—from almost zero thirty-five years ago to over 40 percent today—primarily on the basis of their products' reliability and durability.

8. Weak, Ineffective Reliability Practices

Adding to these pressing needs are our traditional but hopelessly antediluvian approaches to reliability. Part II of this book devotes a chapter to each of the following weaknesses.

- ❑ The obfuscation of mathematical reliability
- ❑ The cookbook approach to reliability prediction
- ❑ The pathetic inability to estimate and forecast reliability
- ❑ The futility and enormous waste of reliability demonstration

Causes for Poor Reliability

There are general (i.e., business culture) causes, as well as specific product lifecycle causes, for poor reliability.

General Causes

- ❑ Increasing product complexity
- ❑ Engineers pushing "the state of the art" for performance
- ❑ Too many features (i.e., bells and whistles) crowding out reliability

- ❑ More severe environments, field stresses, and interactions
- ❑ Compressed development cycles in the rush to be "first in the market"
- ❑ Product lifecycles that are shorter than design cycle time
- ❑ Rapid product obsolescence
- ❑ Rising customer expectations, especially for reliability
- ❑ Lack of financial incentives/penalties for reliability

Specific Product Lifecycle Causes

- ❑ Reliability is not a specification.
- ❑ Customer specifications are not known and are not solicited. (Engineers believe that they know better than the customer what he needs.)
- ❑ Engineers use the latest (but unproven) parts, rather than those with proven field history.
- ❑ Management issues edicts to cut product costs and speed up development at the expense of field failures.
- ❑ Weak links of design, suppliers, and manufacturers are not "smoked out."
- ❑ Powerful demonstration techniques, such as Multiple Environment Over Stress Tests (MEOST), are not known.
- ❑ There is inadequate field pilot testing.
- ❑ There are confusing installation and/or use instructions.
- ❑ There are no built-in diagnostics.

- There is inadequate data retrieval from the field, especially after the warranty period.
- There is weak or lacking independent failure analysis.

The Reach-Out Objectives of True Reliability

An overall objective of this book is to introduce the reader to a revolutionary reliability technique—MEOST—that can dramatically improve reliability right at the design stage, in hours, instead of waiting months and years to confirm unreliability in the field.

Objectives

Specific objectives follow from this overall objective:

1. To expose the needless complexity and waste of mathematical approaches to reliability
2. To highlight the futility of current practices for reliability prediction, reliability estimation, and reliability demonstration
3. To detail practical and effective steps in achieving reliability levels never attained with other techniques
4. To give management a quantified reliability figure instead of fictional reliability figures, with little confidence of success
5. To ensure that design reliability is not degraded by production workmanship or by the supply chain
6. To speedily evaluate the effectiveness of design changes—in hours instead of an anxious wait of months and years in the hands of the customer

The Overwhelming, Tangible Benefits of True Reliability

The ultimate benefit of our approach to reliability, using MEOST, is to advance and enhance the primary goals of a corporation, which are to:

- ❑ Increase customer loyalty and retention.
- ❑ Vastly improve profits.
- ❑ Add value to all stakeholders—customers, employees, suppliers, and stockholders.

The more specific benefits are multidimensional:

1. Reliability improvements that virtually outlaw unreliability—from 10:1 up to 100:1 improvement, striving for field failure rates below 100 parts per million (0.01 percent) a year
2. Discovery (and correction) of potential field failures at the design stage in hours, instead of months and years in the field
3. Drastic reduction in warranty costs by factors of 10:1 up to 100:1
4. Extension of warranties and guarantees to customers up to seven years and even up to the lifetime of the product
5. Virtual elimination of product liability lawsuits, recalls, and retrofits
6. Reverse engineering through MEOST to assess a competitor's reliability
7. Reduction of product cost—MEOST Stage 8

8. Reduction of design test cost, space, manpower, and test samples
9. Reduction of customer downtime
10. Enhancement of product resale value
11. Reduction of design time to market
12. Enhanced job excitement among staff
13. Minimum 5 percent to 10 percent increases in customer retention and longevity
14. A profit increase of 50 percent to 150 percent

The Future Direction for Reliability

Based on our consultations with more than 400 companies in thirty-three countries and four continents, we see several trends unfolding:

- ❑ Companies, having achieved reasonable success with quality, are seeking new frontiers, with reliability in their sights.
- ❑ Reliability is now marketable, predictable, and quantifiable in real terms, not in fictitious, mean-time-between-failure (MTBF) figures.
- ❑ Automotive companies are beginning to gain reliability traction, surpassing military reliability.
- ❑ The necessities of product safety, product liability prevention, and keeping clear of burdensome legislation are driving companies more and more toward reliability.
- ❑ More customer companies are introducing "epidemic" reliability clauses, where manufacturers are forced to "belly up" to the huge financial losses of poor reliability.

- At the same time, manufacturers that have achieved reliability breakthroughs are demanding—and receiving—incentive payments from their customers for meeting reliability targets.
- Reliability is moving from failure percentages and MTBF to failure parts per million (PPM) per year and to FITS (failures in time), where one fit is defined as one failure per billion hours of use.

In short, we see our way clear to being able to achieve dramatic improvements in reliability and to do so economically—what we call "space-age reliability at commercial prices."

An Organizational Infrastructure for World-Class Reliability

Before we begin our reliability journey to the promised land, it is important to design an infrastructure for reliability. There are three essentials:

1. A leadership philosophy and commitment
2. A supportive organizational framework
3. A comprehensive reliability assurance system

A Leadership Philosophy and Commitment

It is well known that no major initiative in a company moves forward without the support and involvement of top management. This means, first, there must be some fundamental beliefs about reliability. Then, there must be a translation of those beliefs into action.

Fundamental Leadership Beliefs

- ❑ Reliability is an important road to the twin objectives of any company—customer loyalty and breakthrough profits.
- ❑ Failures are not inevitable. They can be rooted out.
- ❑ Failure detection and correction can, and must, be achieved in design—there can be no waiting for failures to accumulate at customer sites.

Translating Leadership Beliefs into Action

- ❑ Quantification and measurement of reliability must be done as early as possible.
- ❑ Firm reliability goals must be established.
- ❑ Financial incentives/penalties with key customers and key suppliers must be created.
- ❑ A revolutionary approach to reliability in the design function—MEOST—must be mounted.
- ❑ Key suppliers must be helped in order to achieve specified reliability goals.
- ❑ Resources in manpower and capital equipment for MEOST must be committed up-front.

A Supportive Organizational Framework

Many reliability voyages have been shipwrecked on the rocks of a bad organizational structure. Several organizational principles can prevent such disasters.

- ❑ Reliability is primarily a *design* responsibility, not a quality function.

- The quality/reliability function should be the coach/consultant for design, manufacturing, and the supplier. It should report to top management.
- The independence of the quality/reliability function must be safeguarded.
- The quality/reliability function must have authority to shut down a line or a vendor if quality/reliability is jeopardized, regardless of pressures to ship product.
- Where possible, such shutdown authority should be extended to line operators.
- Independent failure analysis capability, down to root cause, must be established within the company. It should not be at the mercy of suppliers.

A Comprehensive Reliability Assurance System

Reliability assurance entails a holistic approach, involving:

- All stakeholders—customers, leadership, organization, employees, and suppliers/distribution channels
- Key disciplines within quality/reliability, cost, and cycle time
- Key operations—design, manufacturing, and business processes

It's beyond the scope of this book to detail this holistic approach. The reader is encouraged to use *The Power of Ultimate Six Sigma* for a full treatment of this subject.[1] Nevertheless, parts of this book describe specific elements of a reliability assurance system.

Part II discusses what *not to do* regarding reliability predictions, estimators, and demonstrations.

Part III looks at what *must be done* in design, with suppliers and, above all, with the powerful technique of MEOST.

Let us, therefore, begin our long journey to achieve space-age reliability at commercial prices!

PART II

What Not to Do: Current but Ineffective Reliability Methodologies

CHAPTER 3

Reliability Mathematics—Complex and Ineffective

"If mathematical formulas could be invoked, the scourge of field failures would have been eradicated, as completely as smallpox has been almost wiped off the face of the earth."

—KEKI R. BHOTE

The Bathtub Curve Does Not Hold Water!

Reliability is often expressed as failure rate over time and depicted graphically. The most common depiction is the bathtub curve (so-named because it resembles the cross-section of a bathtub). Figure 3-1 shows a bathtub curve (the line labeled A), where it is compared against other failure distributions.

The failure rate at zero time (i.e., when a product is first used) is quite high. It quickly declines in the field within the first three to six months. This period is called the infant mortality period. It then, supposedly, settles down to a relatively

Figure 3-1. Typical failure rates vs. time.

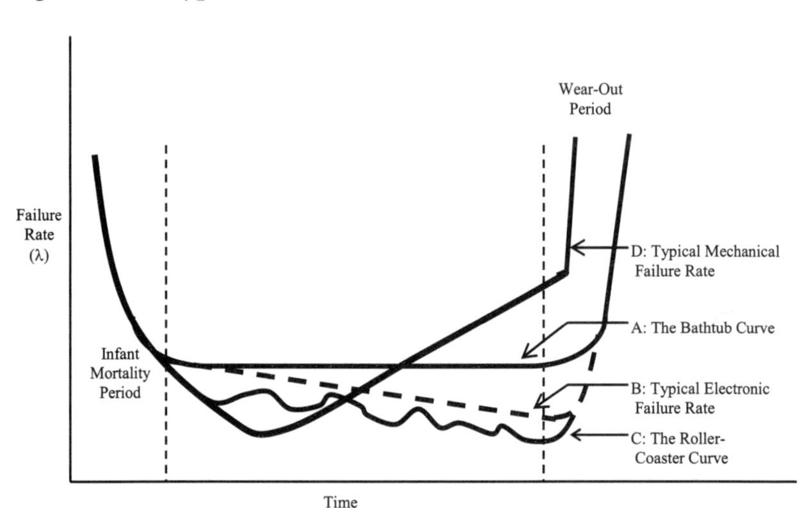

low and constant failure rate, lasting several years. Finally, there is a sharp rise in the failure rate at the end of product life (called the wear-out period).

Much of reliability mathematical theory is based on a constant failure rate (the flat portion of the bathtub curve). It is the basis of the military Mil-Std.-781 series—Reliability Design Qualification and Production Acceptance Test. Unfortunately, this theory is flawed. There is no such phenomenon as a constant failure rate, just as there is no such thing as a random cause of a defect. The bathtub curve just does not hold water.[14] In many products, especially electronics, the failure rate continues to decline—though at a slower rate than in the infant mortality period. This is shown in Figure 3-1(B). Other products display an up-and-down failure rate, known as the roller-coaster curve, as shown in Figure 3-1(C). Mechanical products typically follow a declining failure rate, followed by a rising failure rate, and eventually a sharp wear-out period, reminiscent of the proverbial one-horse shay.

The Convoluted Logic of Military Testing

The military is notorious for its proliferation of specifications. Former Vice President Al Gore, who chaired a commission to reduce waste in government, ridiculed the twenty-six-page specifications of the Defense Department to test an ashtray. There is a formidable array of military specifications—too numerous to list—on just reliability alone. Among one of the military's silliest recommendations is its approach to testing.

As an example, let's say a product requires a reliability of a maximum of 500 failures per million hours (i.e., a mean time between failures—MTBF—of 2,000 hours). Mil-Std.-781 calls for the shortest test time of three times the MTBF, or 6,000 hours. This would require almost a full year of testing for one product. To speed the testing—so the argument goes—why not test 6,000 units for one hour? Of course, this is an exaggeration, but Mil-Std.-781 allows such a trade-off, so you can increase the number of units and shorten the time to demonstrate the same MTBF. If 3 units are tested for 2,000 hours, or 10 units for 600 hours, two serious problems arise. First, even 600 hours is an eternity. (In our MEOST methodology, eight to twenty-four hours would suffice.) Second, the shorter hours would tend to give an artificially higher failure rate, as in Figure 3-1(A).

The Futility of Reliability Mathematics

Reliability testing based on mathematics takes an inordinately long time, incurs a huge cost, and produces dubious results. Other than that, it is pretty good! It is based on several assumptions.

Assumption 1: Reliability Growth

Reliability growth (i.e., improvements) is modeled as a straight line on a log-time versus log-reliability plot that's generally expressed as mean time between failure (MTBF).

Figure 3-2 illustrates that to improve reliability, more test hours must be accumulated. Hundreds of thousands of test hours are not only expensive, they are also unnecessary to reach today's stringent reliability goals. Furthermore, the reliability goal is associated with varying degrees of confidence, generally ranging from 50 percent to 95 percent.

Assumption 2: Binomial Probabilities

The binomial probability distribution is used to calculate the number of test units required to detect one specific failure

Figure 3-2. The reliability growth curve.

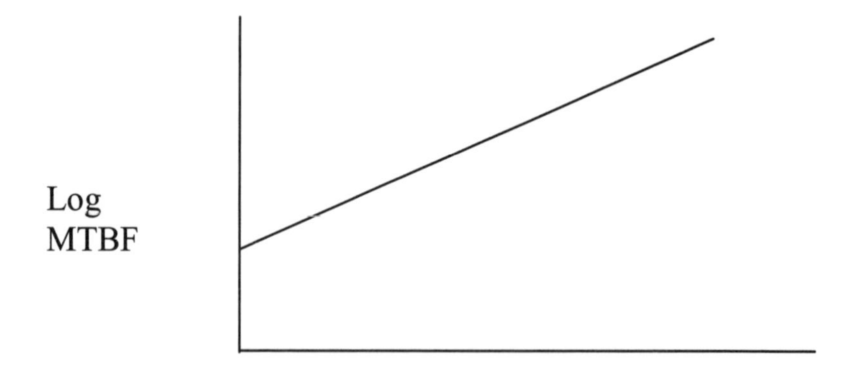

mode. The binomial probability distribution has been converted from its complex mathematical equation to the simpler graphical nomograph of Figure 3-3, with built-in confidence levels.

To use Figure 3-3, select the specified failure rate—say, 2 percent (expressed as a probability of 0.02). Draw a vertical line from the 0.02 location on the X-axis until it intersects a diagonal confidence level—say, 95 percent—and read off the number of units required for the reliability test on the Y-axis. In this example, the required sample would be 160 units.

Figure 3-3. Graphical portrayal of failure rate probabilities and the minimum sample sizes required for various confidence levels.

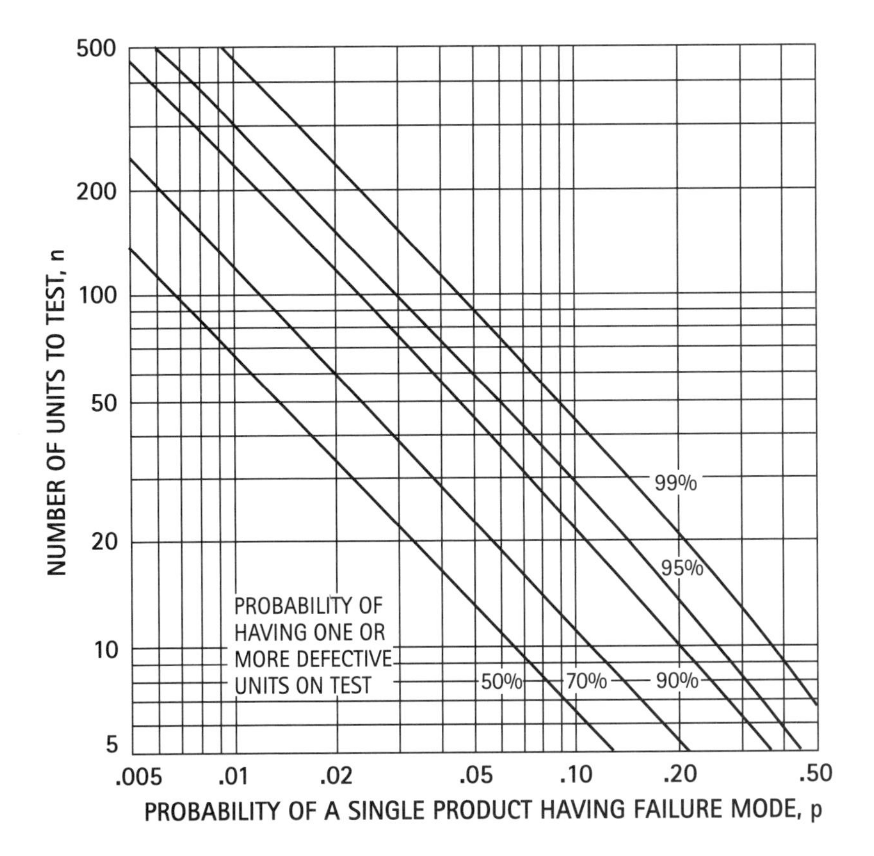

Table 3-1 reduces Figure 3-3 to an easier matrix, with the failure rate percentages—and their reciprocal reliability percentages—on one leg of the matrix *and* the confidence levels on the other, along with required sample sizes for each combination of failure rates and confidence levels.

Table 3-1. Required sample sizes for varying failure rates and varying confidence levels.

Failure	*Reliability*	*Confidence Levels*				
Rate %	*%*	*50%*	*70%*	*90%*	*95%*	*99%*
0.5%	99.5%	145	229	490	570	780
1.0%	99.0%	68	130	225	300	575
2.0%	98.0%	33	58	130	160	230
5.0%	95.0%	13	23	46	59	90
10.0%	90.0%	7	12	23	29	95

Fundamental Weaknesses of the Mathematical Approach

Figure 3-3 and Table 3-1 dramatically highlight the weakness of the mathematical approach to reliability testing.

Weakness 1. Jumbo-Size Quantities Are Required for Reliability Tests. The sample sizes are huge, generally over 100 (for a low reliability level of 99 percent and a significant confidence level of 90 percent or greater). For a modest reliability level of 99.5 percent (i.e., a failure rate of 0.5 percent and a significant confidence level of 90 percent), the quantity required for test jumps to 490 units.

Weakness 2. Costs Are Astronomical. Imagine the cost to build, test, and analyze 490 units at the design stage of a product, when it is realized that not more than 10 units are typically available at the prototype stage of new product design. (Compare this with MEOST sample sizes of 3 to 10 units and their negligible costs.)

Weakness 3. Test Time Is Inordinately Long. The time required to test these units could stretch into months. In the meantime, the design team either has to wait for the results or jump the gun and release the product for production and shipment to the customer prematurely. Rolling the dice in Las Vegas would have greater probability of success! Furthermore, if the tests fail, another round of several months would be required.

Weakness 4. Twentieth-Century Reliability Levels Are Totally Inadequate for the Twenty-First Century. Most of the figures and charts, extrapolated from the binomial distribution, do not go above a reliability level of 99.5 percent. That is a failure rate of 0.5 percent or 5,000 parts per million (PPM) when today, levels below 100 PPM are required. The quantities, test costs, and test times would be astronomical, difficult even to visualize.

Weakness 5. Confidence Levels Below 90 Percent Are Statistically Irrelevant. A 50 percent confidence level is meaningless. It is the probability of tossing a coin for heads or tails. A 70 percent confidence level is not a whole lot better. It is only when confidence levels reach 90 percent or higher that the results are statistically significant.

Weakness 6. Sample Sizes Escalate Beyond a Single Failure Mode. Figure 3-3 and Table 3-1 are valid for only a single failure mode. Doubling the failure modes will triple the sample sizes required, complicating an already-impossible scenario.

Weakness 7. The Poverty of Results. In the final analysis, if meaningful reliability results would accrue from the mathematical approaches, the very high costs and high cycle times may be tolerated. However, the results are still a beautiful piece of fiction. Why? Because the tests are generally conducted in benign, room-ambient conditions with none of the hostile stresses or environments that would be prevalent in actual field conditions.

The Computer Approach to Reliability

For the last twenty years, computer solutions for quality and reliability programs have become popular. They are based on complex equations that can be programmed in the computer. This approach is, at best, a backward look at real-life situations because it uses data from the past to arrive at the mathematical equations. It is the equivalent of trying to steer a car by only looking at the rear-view mirror.

Another weakness is the interactions present among various stresses. Further, the mathematical equation can only handle known stresses for one product and its associated stresses/environments. It would not fit another product with a different set of stresses/environments.

At the very least, a computer simulation should always be verified by an actual Multiple Environment Over Stress Test.

Reliability Failure Distributions

There are several distributions of failure rates over time that have been used to fit a variety of reliability applications. Table 3-2 lists these distributions and their associated features.

Of these distributions, the well-known normal distribution, widely used in quality work, is not applicable in reliability because failures are seldom, if ever, normally distributed. Similarly, the exponential distribution, which assumes a constant failure rate in the flat portion of the bathtub curve, does not happen in real life. Instead, as Figure 3-1 shows, there is a declining failure rate for electronics products, a roller-coaster

Table 3-2. Reliability distributions and their features.

Distribution	*Features*
Normal	• Many types of failure modes are normally distributed. • Based on the Central Limit Theorem, they are described mathematically by a mean (or average) "μ" and a standard deviation "σ." • Batteries and incandescent lamps are examples of products that follow this distribution.
Log Normal	• Similar to normal distribution, except that linear time is replaced by log time. • It resembles a "skewed" normal curve with a left or right tail. • Many semiconductors, propagation of cracks, maintenance activities, or degradation failure modes appear to have a log normal distribution.
Exponential	• Constant failure rate (the low flat portion of the reliability bathtub curve). Exponential distribution is widely used in reliability predictions, but is nonexistent in actual practice.
Gamma	• Occurs when partial failures are allowed. It is primarily used with redundant systems.
Weibull	• This is the only distribution of importance to practical reliability users, and it is a powerful reliability tool. • Its shape is formed by three parameters that can fit any type of failure distribution.

failure rate for other products, and a slow rising failure rate in mechanical products in the so-called flat portion of the bathtub curve.

The Weibull Distribution

Of all the failure distributions, the Weibull distribution is the only one that fits any type of failure distribution. It easily models failures associated with bearings, semiconductors, capacitors, electronic systems, mechanical systems, pumps, and materials. Its ability to do so rests on three Weibull parameters: α, β, and γ.

α is the scaling parameter, or the time when 63.2 percent of the original units have failed. It is also called the characteristic life and is similar to the mean of other distributions.

β is the shaping parameter that can be used to fit many different reliability failure distributions. This is the slope on the Weibull graphical plot.

γ is an offset in life, referred to as a failure free time. When the number for γ is positive, it represents a time prior to first failure.

Of these parameters, β—or slope of the plot—is most useful in determining failure rate situations. For instance:

$\beta = < 1$ is a decreasing failure rate. This often occurs in semiconductor-dominated systems.

$\beta = 1$ is a constant failure rate system. This rarely occurs on complex systems.

$\beta = 2$ is a linearly increasing failure rate.

$\beta = 3.4$–3.5 is typical of a normal distribution.

$\beta = \geq 2$ is typical of wear-out. The larger the number, the faster the wear-out.

Graphical Solutions Using Weibull Plots

Fortunately, complex mathematical formulas associated with the Weibull distribution have been reduced to simple graphical plots by Professor Kao of Cornell University. The vertical scale is a $\log_e$, $\log_e$ of the mathematical formula of the failure measurement. The horizontal is a $\log_e$ of the time measurement.

Weibull graphical paper can record failure measurements

as a percent of failures versus time. These cumulative failure percentages can then be fitted through with a best-fit straight line, where slope, β, will determine the likely failure distribution, as described previously.

Weibull analysis is best illustrated by an example. Three types of failure data had 19 failures each, as indicated in Table 3-3. Had the data been plotted as failures versus linear hours to failure, Column I would have shown a rapidly decreasing failure rate (infant mortality), Column II a random but somewhat constant failure rate, and Column III a normal distribution, suggesting a wear-out period.

Figure 3-4 shows the same data on a Weibull probability graphical plot (where the X-axis is life in hours on a $\log_e$ scale and the Y-axis is the percentage of failures on a $\log_e$, $\log_e$ scale). The Weibull plot shows all three sets of data as close to *linear*. (This permits just six and sometimes even three data points

Table 3-3. Three separate hours to failure.

I (Hours)	*II (Hours)*	*III (Hours)*
55	422	14,259
25,158	29,144	19,953
7,793	3,566	17,009
5,038	1,805	16,726
2,301	6,127	19,770
9,300	1,330	23,595
5,151	36,003	13,004
15,751	11,100	17,255
2,727	6,600	11,850
117	6,798	13,750
1,432	16,075	15,470
2,427	6,798	18,345
1,510	22,204	18,565
12,290	5,000	15,110
973	10,099	18,144
4,299	24,211	17,706
231	2,058	17,196
1,007	13,002	16,001
3,083	13,564	18,776

Figure 3-4. Weibull plot—hours to failure.

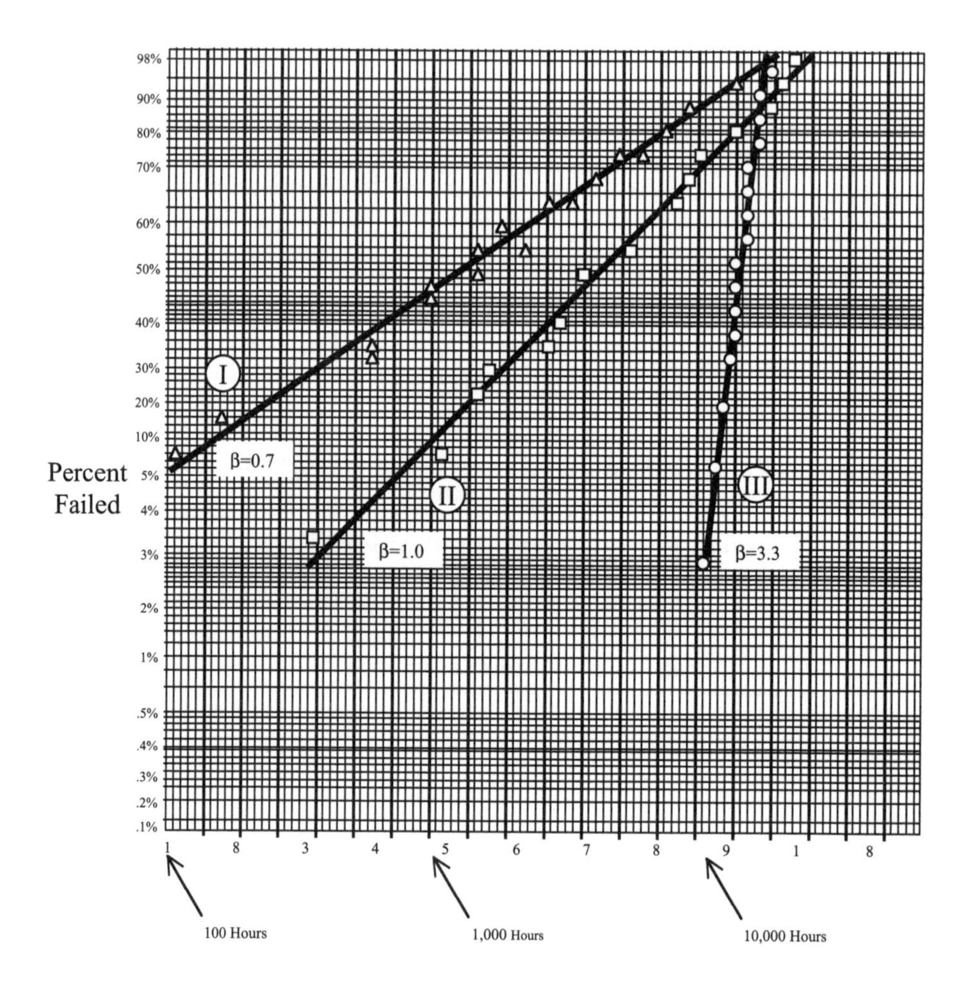

to be joined in a best-fit straight line and extrapolated beyond the shortest failure times.) Column 1 has a β slope of < 1, indicating infant mortality; Column 2 has a β slope close to 1, indicating a constant failure rate; while Column 3 has a β slope of > 3, indicating a wear-out period.

Main Weaknesses of Mathematical Approaches to Reliability

We can now summarize the main weaknesses of a mathematical approach to reliability:

- ❑ The formulas are complex and difficult even for engineers to understand.
- ❑ The constant failure rate model—used in most reliability calculations and projections—is fictitious and flawed.
- ❑ The tests require a prodigious amount of units.
- ❑ The tests take too long—from several hundred hours to several thousand hours.
- ❑ The tests are expensive in terms of the manpower required.
- ❑ The reliability levels reached are too low for today's world.
- ❑ The "unkindest cut of all" is that these tests do next to nothing in rooting out the causes of poor reliability and advancing true reliability growth.

In the next three chapters, we will explore the shortcomings of traditional reliability prediction, reliability estimation, and reliability demonstration.

CHAPTER 4

Reliability Predictors—Cooking the Books with Reliability Cookbooks

"It has become fashionable, in quality, to flood industry with standards upon standards, like ISO-9000, that beget even more standards. They do little to advance companies to world-class quality. . . .

Similarly, in reliability, we are deluged with cookbooks for predicting reliability, but they have as much probability for serving up a winning recipe as a roll of the dice in Las Vegas. . . ."

—KEKI R. BHOTE

A Plethora of Reliability Cookbooks

Just as every state in the Union has developed its own version of the Malcolm Baldrige National Quality Award, countries feel that they cannot achieve reliability heaven without a national guideline for reliability.

The granddaddy of these cookbook approaches to reliability was RCA's *TR1100,* initiated as far back as 1956. It established failure rates for the most frequently used electronic parts. This was soon followed by *Mil-Handbook 217,* now in its e-version, issued by the Department of Defense (DOD) as a reliability prediction bible. Under the sponsorship of DOD, every military contractor is encouraged to use it.

The British issued a similar document—the *British Telecom (BT) Handbook.* The French, who are afflicted with the NIH (not invented here) syndrome, invented their own *French National Center for Telecommunications Handbook,* dubbed *CNET.* And the Japanese, not to be left behind, issued the *Nippon Telegraph and Telephone (NTT)* reliability table. Today, fast-buck software companies have weighed in with several software programs, and they miss reliability accuracy by a mile.

Over 600:1 Variation in Reliability Among the Handbooks

Table 4-1 shows the astonishing differences among the cookbooks—the so-called national standards—in their reliability

Table 4-1. Four national handbook reliability predictions of a large memory board.

	Board-Predicted Failure Rate	
National Handbook	*Fits**	*% Per Year*
Mil-Handbook 217E (U.S.)	4,240,460	371.3
BT (Britain)	700	11.6
CNET (France)	37,870	33.0
NTT (Japan)	37,940	33.1

* 1 Fit = 1 Failure in 1 Billion Hours.

Source: Kam L. Wong, "The Bathtub Curve Does Not Hold Water Anymore," Quality and Reliability Engineering Symposium (1988); and "What Is Wrong with the Existing Reliability Prediction Methods?" *Quality and Reliability Engineering International* (1991).

predictions of the same large memory board.[14,15] The differences are over 600:1. So much for cookbook accuracy! (When we were forced to use *Mil-Handbook 217* thirty years ago, the predicted failure rates of the equipment we were designing were so high that we would have had to commit collective suicide if the predicted failure rates actually occurred.)

Reasons for Inaccuracies Among the Handbooks

Mil-Handbook 217E and the other handbooks do consider part complexity, part technology, package technology, part application, electrical stress, temperature, and manufacturing quality-control level for their failure rate projections. They do this with exponential extrapolations and multipliers. What they do not consider are the following factors:

1. *Temperature Cycling.* Failures accelerate with 1) the number of thermal cycles and 2) the wider ranges from cold to hot. Figure 4-1 shows that failures can increase sevenfold with these two factors alone. In addition, the rate of temperature change—say, from 1° C or 2° C per minute to 25°

Figure 4-1. IC failures vs. number of thermal cycles and temperature swings.

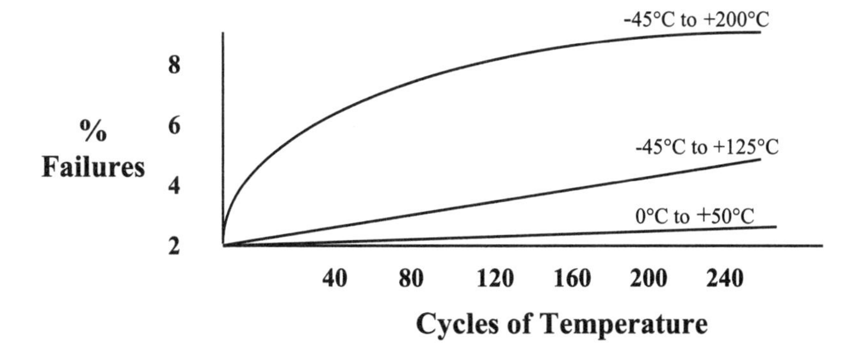

C to 40° C per minute—generates far more rapid failures. These are very important factors in our MEOST tests.

2. *Failure Rate Changes with Materials.* For brittle materials, a higher rate of temperature change accelerates failures, whereas for ductile material such as aluminum and solder, a lower rate of temperature change accelerates failure.
3. *Vibration: Sensitivity to Exact Installation and Mounting Structure.* The amplitude of vibration can vary within a product, depending on the exact installation and the mounting structure. Each unit would have its own response spectrum and, hence, its own failure rate when subjected to the same vibration input.
4. *Nonoperating Failure Rates.* It may come as a surprise that non-operating products/systems are still subject to environmental stresses. These non-operating—or zero stress—failure rates vary under different *storage* and *shipping* conditions. A Rome Air Force Development Center study revealed that nonoperating failures contributed 10 percent to 30 percent of the total number of failures for aircraft with a utilization of twenty to sixty hours per month. It is difficult to separate non-operating failures from first-time turn-on failures. Many missile storage studies have found that failures appear to be independent of the length of storage. This strongly suggests that turn-on was the culprit in most of these failures.
5. *Combined Environments.* While temperature cycling and vibration are the two most important stress/environments causing failures, the handbooks do not take into account other stresses, such as humidity, dust, altitude, power cycling, transients, or radiation. Yet their effect in simultaneous combination cannot be ignored.
6. *Supplier Variations Affecting Reliability.* Handbooks do take into account reliability specifications imposed on suppliers,

such as verification of parts mechanical integrity, long-term measurement failure rates, minimum life expectancy, extent of parametric measurements, and the amount of environmental screening. But not all reliability requirements can be specified; if they were, the costs would be prohibitive. Studies have shown that even qualified suppliers have widely different failure rates. Figure 4-2 indicates a 30:1 difference in percentage failure rates between the best and worst suppliers when an RF inductor was subjected to several hundred thermal cycles. Figure 4-2 also shows how repeated thermal cycles will accelerate failures—one of the principles used in MEOST.

Figure 4-2. RF indicator failure rate from four suppliers, subjected to several hundred thermal cycles from –55° C to +125° C.

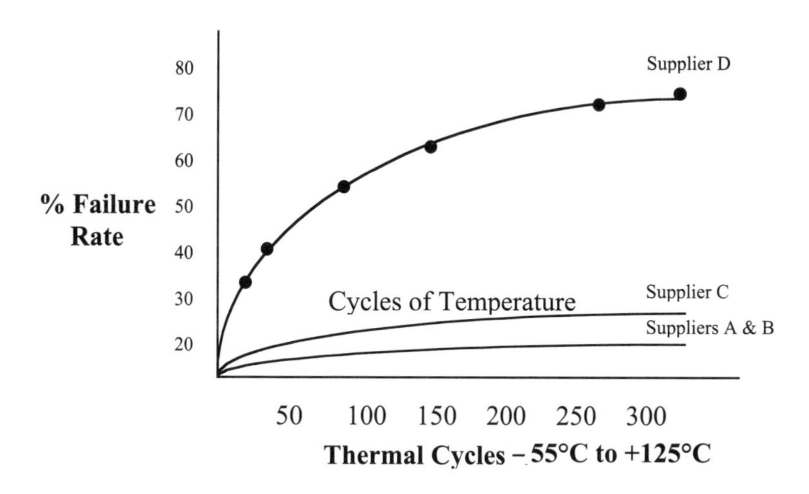

A Company's Own Field History—A Better Predictor

A better approach to reliability prediction is for a company to closely track its own field history within the warranty period and, more important, beyond the warranty period. It should establish a "library" of field failure rates, part by part, taking into account derating (i.e., factors of safety in stress reduction), applications, field environments, and time exposure in the field. Companies will have much greater success in reliability prediction by using this library approach.

There is, however, a caveat. Most companies do not get accurate feedback on their field failures. Among the reasons are lack of information from customers, service writers, and service stations; imprecise field age; customer applications and field environments; customers' lack of knowledge and/or abuse of the equipment; superficial failure analysis; and an almost total lack of failure data after the warranty period. These are more administration problems than reliability problems, but they have to be addressed. One of the best ways to get a handle on out-of-warranty failures is through the use of MEOST, especially stages 6 and 7, as detailed in Chapter 10.

CHAPTER 5

Reliability Estimators—A Cloudy Crystal Ball

> *"Don't listen to the engineers . . . Let the parts do the talking!"*
>
> —DORIAN SHAININ

"Regardless of specifications, the product must work." Those were the words of Fred Hill, a Motorola vice president, in an internal memo to the company's engineers. It is right to emphasize the need to first determine customer requirements before determining specifications, and then to determine specifications before preliminary design. But when it comes to reliability, the customer requirement is simple and straightforward: "The product must work." It must live in the hands of the customer for a specified time, under specified environments—*without failures.*

Failure Mode and Effects Analysis (FMEA)

The FMEA has been the main conventional technique of estimating reliability. It was first developed by Grumman Aircraft in 1956 for aerospace work and by NASA. The Japanese today are great proponents of FMEA. The technique has made its way into Europe; but in the United States, it is hardly known or used, except for the automotive industry. For the U.S., that turns out to be a lucky break.

As a reliability tool, FMEA's main use should be only as an elementary method to pinpoint the weakness of a design in terms of potential failures that are likely to occur. Its primary weakness is that these potential failures are only the opinions, judgments, theories, and hunches of the designers. As such, they are far less reliable than "talking to the parts," as the great Dorian Shainin—one of the world's foremost problem solvers—has attested and proved. In quality problems, "talking to the parts" with simple but powerful Design of Experiments (DOE) using techniques[3] designed by Shainin and Bhote is far more effective and statistically powerful than the complex, costly, and statistically weak techniques practiced by Six Sigma consultants and their black belt followers.

In reliability, the way to "talk to the parts" is with Multiple Environment Over Stress Tests (MEOST), a technique detailed in Part III of this book. MEOST is far more effective than engineering guesses of possible failures, which is what you get using the FMEA methodology. Hence, only a brief discussion of FMEA will suffice.

FMEA Methodology

As shown in Table 5-1, the start (first column) of an FMEA is a list of the most likely failures, usually at the component/

part level, along with a description—in one or two words—of those components. In the second column is, again, a one- or two-word description of the failure mode. The third column is a brief description of the effect of that failure upon the customer. The fourth column lists the causes of that failure.

Next are three columns that assess the impact of that failure, *in the designer's opinion*. The first (i.e., the fifth column) is the probability of that failure in the field rated on a subjective scale of 1–10, with 1 being the least probable and 10 indicating the highest probability of failure. The next (i.e., sixth column) is the severity of that failure on the customer—again, expressed as the designer's opinion and rated on the same scale of 1–10. The next (i.e., seventh column) is the probability of nondetection of such a failure in the plant through inspection/test. Once again, the same 1–10 scale is used, with 1 being a failure that's easily detected in the plant and 10 indicating a failure that's highly nondetectable. The next column (i.e., the eighth one)—called the risk priority number—is the multiplication of the probability, severity, and detectability numbers. The highest risk for any part is 1,000 (i.e., 10 × 10 × 10), which merits the highest remedial action. A lower risk, such as 3–10, indicates that the reliability concern associated with that part is low, and no action is necessary.

Next follows the most important part of an FMEA exercise—what corrective action to take (column nine). This action can be 1) changing the supplier, 2) changing the design, 3) adding a process step, 4) adding an extra test, or 5) creating redundancy (usually a military option).

As a result of the proposed corrective action, the three assessments—probability of failure, severity of failure, and probability of nondetection in the plant—are reassessed. These three numbers, multiplied together, should now give a much lower risk priority number (RPN) than was recorded in column eight. This is how reliability is enhanced in an FMEA.

Table 5-1 is an example of an FMEA conducted on a Ford electronic system. It is just one page selected from several pages of the FMEA. One of the most critical components in this FMEA was the 400 NH Toroidal Inductor. It displayed a vibration defect, resulting in the highest risk priority number of 900. The corrective action was to epoxy the toroid to the circuit board to prevent vibration failure, thereby reducing the RPN from 900 to 9.

Limitations of FMEA

There are several weaknesses associated with an FMEA study:

- Its dominant weakness is that the selection of the most likely components to fail is based on the engineer/designer's opinion, rather than the factual certainty of "talking to the parts" (the method used in the MEOST methodology). History has shown that an FMEA rarely replicates the actual and subsequent field failures on the product.
- It is only a preliminary paper study—best performed at the pre-prototype stage of design. Too much time should never be wasted on an FMEA. (Many customers, especially automotive companies, insist on every single component being analyzed with an FMEA, which can run into page after page of make-work. Apparently, these companies evaluate the effectiveness of an FMEA by the weight, in pounds, of the report.)
- It does not address the necessity to truly determine customer needs and requirements (performed with techniques such as

Table 5-1. Failure mode and effects analysis (FMEA).

Project Name: Ford DIS
Power Supply (Page 1)

Part Name/ Function	Failure Mode	Effect of Failure	Cause(s) of Failure	Occurrence	Severity	Detection	Priority	Recommended Corrective Actions and Status	Occurrence	Severity	Detection	Priority
C3	Open	Decreased noise filtering of battery line	Defect in					Will be changed to 21 MF by development engineering for improved filter action (issue '0')				
600 uF/35V Electrolytic Capacitor			Component manufacture, excessive vibration	1	2	10	20		1	2	10	20
F1 - Filter	Short	Catastrophic. Fuse F1 blows. ATC enters default defrost mode. DIS inoperative.	Excessive heat, voltage transient defect in component manufacture	5	9	1	45		5	9	1	45
L1	Open	Catastrophic. ATC enters	Excessive					Epoxy toroid to board to prevent vibration failure				

(continues)

Table 5-1. (continued)

400 NH Toroidal Inductor		Vibration defect in component manufacture	Vibration defect in component manufacture	10	9	10	900		1	9	9	9
P1 - Filter	Short	Decreased noise filtering of battery line	Excessive					Will be changed to 47 MF by development engineering for improved filter action (issue '0')				
660 MF/35V Electrolytic Capacitor			Component manufacture excessive vibration	1	2	10	20		1	2	10	20
P1 - Filter	Short	Catastrophic. Fuse F1 blows. ATC enters default defrost mode. DIS inoperative.	Excessive									
			Heat, voltage transient defect in component manufacture	5	9	1	45		5	9	1	45
C19	Open	Decreased stability of keep-alive circuit for clock microprocessor.	Defect in component manufacture					Has been changed to 100 MF for stability of keep-alive circuit at +80C (issue '0')				
1 MF Monolithic Capacitor				5	9	10	450		5	9	1	45
U27 Out-put	Short	Catastrophic. No keep-alive voltage for clock microprocessor. ATC enters default defrost mode. DIS inoperative.	Defect in component manufacture									
				5	9	10	450		5	9	1	45

mass customization and quality function deployment) or to translate such requirements into engineering specifications before the start of the design.

Precautions in Using FMEA

1. Keep the list of potentially unreliable components as short as possible. Use only components with proven, good field history.
2. Because suppliers and processes can create field failures, if not controlled, extend FMEAs beyond Design FMEAs to:
   a. Key Supplier FMEAs
   b. Process/Equipment FMEAs
3. Concentrate on the "corrective action" section of the FMEA, since it is the most important segment in reliability improvement.

Fault Tree Analysis (FTA)

There is a companion piece to an FMEA called the fault tree analysis (FTA). Whereas an FMEA starts with the causes of a failure and moves to its effect on the customer, an FTA starts with the effect of a failure on the customer and systematically progresses to the possible causes and eventually down to the root cause.

Table 5-2 is an example of an FTA on an ignition amplifier on an automobile. An FTA is a powerful tool in troubleshooting for field service and the service industry, but it is of marginal value for reliability.

Table 5-2. Fault tree analysis (FTA).

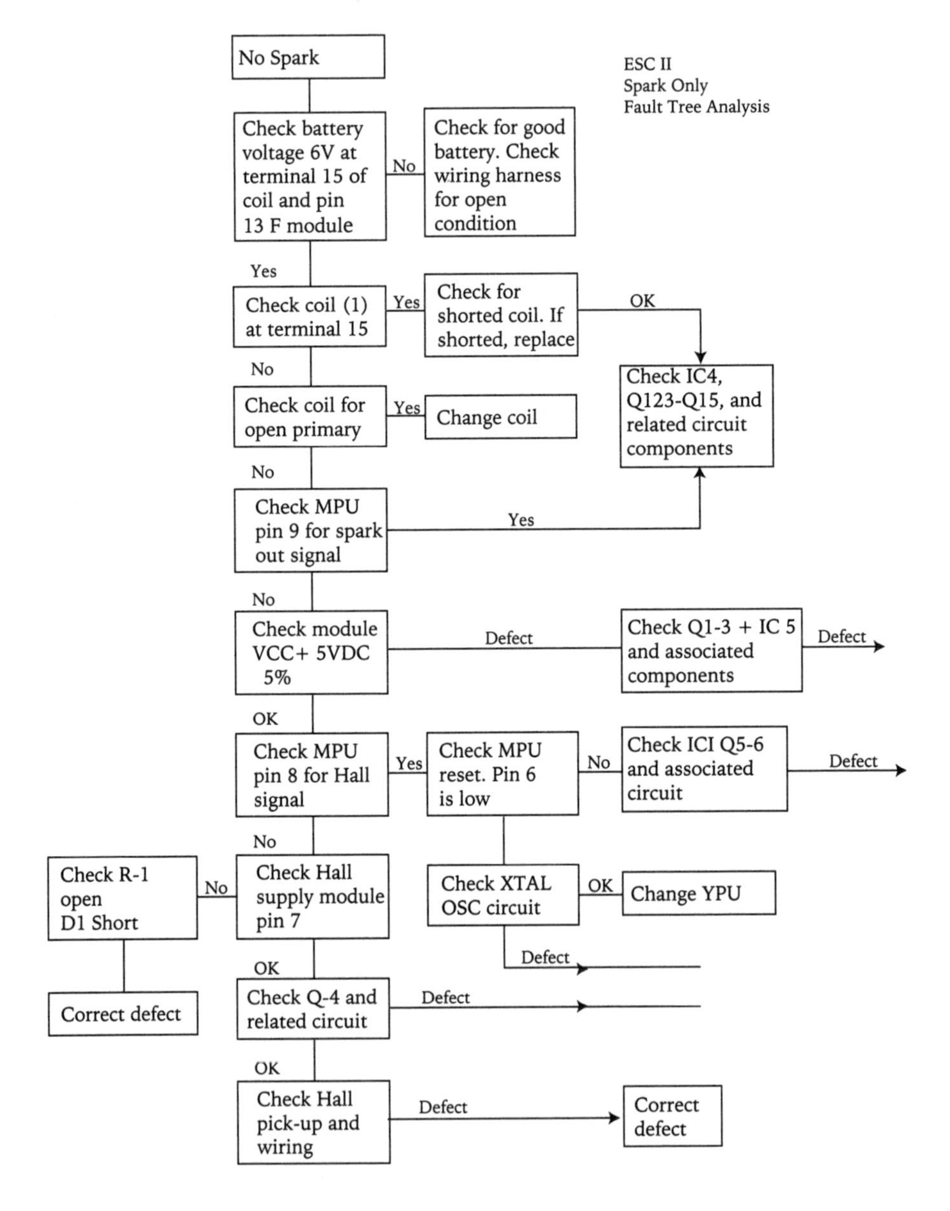

Conclusion

Reliability estimation, using the FMEA discipline, is a little better than mathematical reliability prediction, outlined in the previous chapter. It is, at best, a kindergarten discipline and should be used, sparingly, at the feasibility phase of a design.

We now go to the third leg of a wobbly triumvirate—namely, traditional reliability demonstration.

CHAPTER 6

Reliability Demonstration—Throwing Money at the Problem

"He who knows not and knows not he knows not, he is a fool. Shun him. . . ."

—OLD PROVERB

An Evolution in Reliability Demonstration

It has taken half a century for the quality discipline to evolve. After a hesitant start with statistical quality control (SQC) and sampling plans, through the tentative stages of zero defects, quality circles, statistical process control (SPC), and total quality management (TQM), quality finally reached its pedestal of Six Sigma. It has taken the reliability discipline almost as long to evolve from lifeless life tests to 100 percent burn-in tests (that should be burnt out!) to more productive thermal cycling and vibration testing to HALT and HASS and, finally, reaching its pinnacle in MEOST.

Life Testing

In the early life testing of the 1960s, a sample of products was simply allowed to "cook" for 1,000 to 3,000 hours to detect failures. As expected, few failures were uncovered.

Burn-In: If You Don't Know What Else to Do

The next step was to subject units (especially electronic products) to a high temperature soak for twenty-four to ninety-six hours, on a 100 percent basis. Minor tinkering around the edges added electrical power to the units subjected to burn-in, along with cycling that power "on" and "off." Strangely, this is still the preferred method of demonstrating reliability. Even more incredibly, many customers insist on burn-in as proof of reliability.

What is wrong with 100 percent burn-in? There are several flaws to this technique:

- A 100 percent test, instead of a sample test, is statistically dumb. We know that 100 percent inspection is not 100 percent perfect. (The next time a believer in 100 percent testing goes to his doctor for a blood test, he should ask for 100 percent of his blood to be drawn instead of just a sample pinprick!)
- A 100 percent burn-in adds to manufacturing cycle time. These days, when companies strive to reduce their manufacturing cycle time from days to hours, and a few from hours to minutes, adding twenty-four to ninety-six hours of cycle time in burn-in is even more dumb.
- Most companies do not have a clue as to how to demonstrate reliability, so they latch on to 100 percent burn-in as a Band-Aid.

- Finally, its most serious weakness is that it rarely weeds out even 5 percent of the potential failures in the field.

The Military Boondoggle

The military track record of over-specifying and under-performing is well known in reliability circles. It is based on the same mathematics of reliability, detailed in Chapter 3, and characterized by large quantities specified for that test, long test durations, high costs, inadequate confidence levels, and limited stress levels.

At quality/reliability conferences, Defense Department spokesmen bristle when told that automotive reliability has far surpassed military reliability. Yet the truth of the matter is that 60 percent to 75 percent of military hardware is out of commission at any one time, and the costs of military maintenance are multiples of its procurement cost. The long list of national setbacks includes:

- The scandalous unreliability of U.S. equipment that aborted President Carter's bold mission to rescue our American hostages in Iran in 1979.
- The *Challenger* disaster of 1986, resulting from an unreliable O-ring.

We must, regretfully, add the tragedy of the *Columbia* shuttle breaking up in space and killing seven heroic astronauts because of a chunk of foam, acting as a missile and smashing into the shuttle's left wing.

The scathing report of the commission investigating the *Columbia* disaster was highly critical of NASA's preoccupation with cost and internal politics at the expense of safety reliability. The commission concluded that "if these persistent system flaws are not resolved, the scene is set for another major accident."

Thermal Cycling: The Dawn of True Reliability Progress

In the mid-1960s, the principle of exercising stress on parts and products through temperature cycling was introduced. Within a decade, thermal cycling advanced in seven stages, as shown in Table 6-1. Field reliability improved, but, as we say in mathematics, thermal cycling proved to be a necessary but not a sufficient step. In hindsight, the reasons were that 1) stress/environments were not combined to produce interactive failures; 2) overstress was not extended far enough; and 3) the rate of stress increase was not fast enough. These three principles were later incorporated in Multiple Environment Over Stress Tests (MEOST).

Vibration: A Companion Development

Parallel to thermal cycling, advances were also made in vibration to simulate product failures, as shown in Table 6-2.

Table 6-1. The evolution of thermal cycling.

Stage	*Specific Technique*
1	Thermal Cycling, No Electrical Power
2	Thermal Cycling, Continuous Power
3	Thermal Cycling, Interrupted Power
4	Stage 3: From 0° C to +50° C, 1 Cycle
5	Stage 4: 5 Cycles with Measurements at Temperature Extremes
6	Stage 5: Extension: From –30° C to +85° C, 25 Cycle
7	Stage 6: Extension to > 100 Cycles

Table 6-2. The evolution of vibration testing.

Stage	*Specific Technique*
1	Sinusoidal
2	Single Axis, Single Frequency
3	Sine Sweep
4	Random
5	Random with 6 degrees of freedom (in 3 distinct axes and 3 rotational axes simultaneously)

Other Stresses/Environments

Even though thermal cycling and vibration are considered to be the two most important stresses in simulating failures, others should be considered as well.

Thermal Shock

The difference between thermal cycling and thermal shock is the rate of change from "hot" to "cold" and vice versa. In the old days, the rate of thermal change in temperature cycling was 2° C to 5° C/minute—a pace considered much too slow to exercise failures. Today, thermal cycling ramp-up and ramp-down rates are 25° C/minute, and even up to 60° C/minute.

In thermal shock, on the other hand, the rate of temperature change is almost instantaneous—from 150° C to –50° C in seconds. Semiconductor manufacturers do not recommend thermal shock. It actually reduces product life (although some companies find it applicable for forcing integrated circuit failures). Another disadvantage is the removal of power while the product is being transferred from hot to cold.

Humidity

Exposing products to humidity precipitates corrosion and contamination defects. Humidity can penetrate porous materials, causing leakage and electrolysis. It is an important stress in evaluating coatings and seals. In addition, because oxygen is required for oxidation, humidity limits the uses of nitrogen-cooled chambers. Nitrogen creates an inert environment that inhibits corrosion. Typically, humidity tests are of long duration. Furthermore, certain types of corrosion require minimal airflow, contrary to most chamber usage.

HAST (Highly Accelerated Stress Testing)

This is another type of humidity testing, done in a pressurized chamber and analogous to a pressure cooker or autoclave. HAST aggressively forces moisture in potential failure areas.

Power Cycling and Voltage Margining

Power cycling and voltage margining are two types of electrical stress. Power cycling turns the product "on" and "off" for a number of cycles. Voltage margining varies the input voltage above and below nominal levels. A subset of voltage margining is *frequency margining*. These electrical stresses are usually not sufficient stimuli to precipitate failures by themselves. Their greatest use is in combination with other stresses—as in MEOST.

Figure 6-1 is one chamber company's interpretation of the magnitude of each of several stresses and their possible interaction effects on failure precipitation. The real magnitude and interactions are far more complex.

Table 6-3 is a more comprehensive tabulation[17] of various stresses, their failure effects on associated components/parts and suggested reliability improvement techniques.

Figure 6-1. A crude approximation of the relative influences of various stresses and their actions.

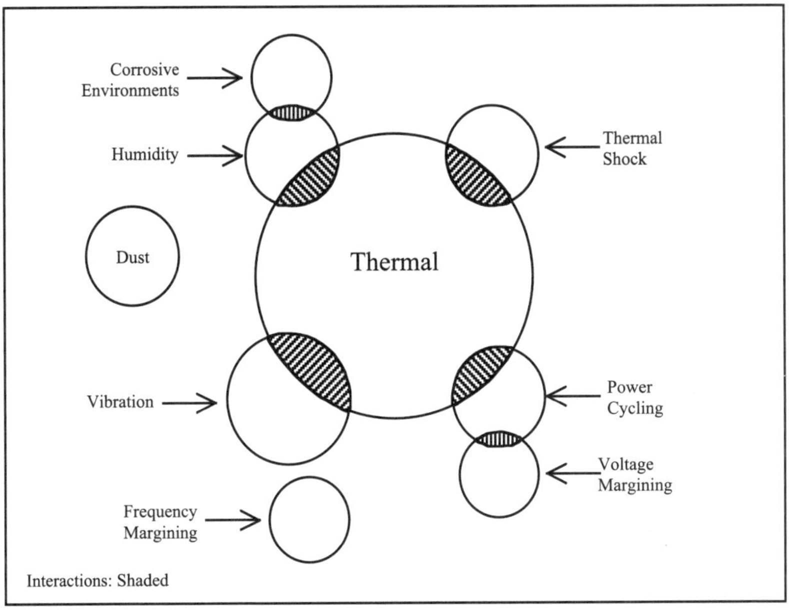

Source: Fundamentals of Accelerated Stress Testing (Holland, MI: Thermotron Industries, 1998).

Failures Precipitated by Thermal Cycling and Vibration

Table 6-4 is a list of the types of failures detected by thermal cycling and by vibration, respectively.[18] It must be recognized, however, that all these failures, researched and reported, are the result of a single environmental stress and, as such, do not stimulate *interaction effects*—a major weakness of any single stress.

(text continues on page 84)

Table 6-3. Environmental stresses, effects, and potential reliability improvement techniques.

Environmental Stress	*Effects*	*Potential Reliability Improvement Techniques*
High Temperature	Parameters such as resistance, inductance, capacitance, power, and dielectric constant will vary; insulation may soften; moving parts may jam due to expansion; finish may blister; thermal aging, oxidation, and other chemical reactions may be enhanced; viscosity may be reduced and evaporation of lubricants can arise; structural overloads may occur due to physical expansion.	Thermal insulation, heat-withstanding materials, cooling systems
Low Temperature	Plastics and rubbers lose flexibility and become brittle; electrical constants vary; ice formation occurs when moisture is present; lubricants and gels increase viscosity; finishes may crack; structures may be overloaded due to physical contraction.	Thermal insulation, cold-withstanding materials, cooling systems
Thermal Cycling and Shock	Materials may be instantaneously overstressed, causing cracks and mechanical failures; electrical properties may be permanently altered; crazing delamination, ruptured seals can arise.	Combination of techniques for high and low temperatures
Shock	Mechanical structures may be overstressed, causing weakening or collapse; items may be ripped from their mounts; mechanical functions may be impaired.	Strengthened structural members, reduced inertia and moment, shock-absorbing mounts
Vibration	Mechanical strength may deteriorate due to fatigue or overstress; electrical signals may be erroneously modulated; materials and structure may be cracked, displaced, or shaken loose from mounts; mechanical functions may be impaired.	Stiffening control of resonance
Humidity	Penetrates porous substances and causes leakage paths between electrical conductors; causes oxidation, which may lead to	Moisture-resistant materials, dehumidifiers, protective coatings, hermetic sealing

Environmental Stress	*Effects*	*Potential Reliability Improvement Techniques*
	corrosion; moisture causes swelling in materials such as gaskets; excessive loss of humidity can cause embrittlement.	
Contaminated Atmosphere Spray	Many contaminants combined with water provide a good conductor that can lower insulation resistance; causes galvanic corrosion of metals and accelerates chemical corrosion.	Nonmetal product covers, reduced use of dissimilar metals in contact, hermetic sealing, dehumidifiers
Electromagnetic Radiation	Causes spurious and erroneous signals from electrical and electronic equipment and components; may cause complete disruption of normal electronic equipment, such as communication and measuring systems.	Shielding, radiation hardening
Nuclear/Cosmetic Radiation	Causes heating and thermal aging; can alter chemical, physical, and electrical properties of materials; can produce gases and secondary radiation; can cause oxidation and discoloration of surfaces; damages electrical and electronic components, especially semi-conductors.	Shielding, radiation hardening
Sand and Dust	Finely finished surfaces are scratched and abraded; friction between surfaces may be increased; lubricants can be contaminated; orifices may become clogged; materials may be worn, cracked, or chipped; abrasions, contaminates insulation, corona paths.	Air-filtering, wear-proof materials, sealing
Low Pressure (High Altitude)	Structures such as containers and tanks are overstressed and can be exploded or fractured; seals may leak; air bubbles in materials may increase due to lack of cooling medium; insulation may suffer arcing breakdown; ozone may be formed; outgassing is more likely.	Increased mechanical strength of containers, pressurization, alternate liquids (low volatility), improved insulation, improved heat transfer methods

Source: CALCE Electronic Packaging Research Center, University of Maryland, May 15, 1995.

Table 6-4. Failure types detected during environmental stress.

	Environmental Stress	
Failure Type	*Thermal Cycling*	*Vibration*
Parameter Drift	X	
Intermittent	X	X
Poor Bond		X
Poor Die Attachment		X
Film Resistor Metalization	X	
Metalization Migration	X	X
Contaminants	X	X
Poor Connections		X
Insufficient/Cold Solder	X	X
Broken/Weak Lead		X
Chafed/Pinched/Loose Wires		X
Defective Lead Weld		X
Loose Parts/Fasteners/Hardware		X
PCB Shorts/Opens	X	X
Defective Parts	X	X
Improper Torque		X

Single Environment Sequential Tests: Another False Start

In an attempt to achieve better reliability, some automotive and electronic companies initiated a long and tortured test regimen, wherein a product would be subjected to a series of separate, single environmental stresses, *but in sequence*. A typical sequence would be: test; thermal cycling; retest, vibration, retest, humidity, retest, etc.

The objective of single environmental sequential testing was to generate failure modes from each test. A second objective was to find these failure modes in months of testing versus years in the field. Figure 6-2 is an example. A pager product was subjected to two months of testing to simulate five years of accumulated failures in the field. The failures were

Figure 6-2. Example of single environmental sequential testing on a radio pager.

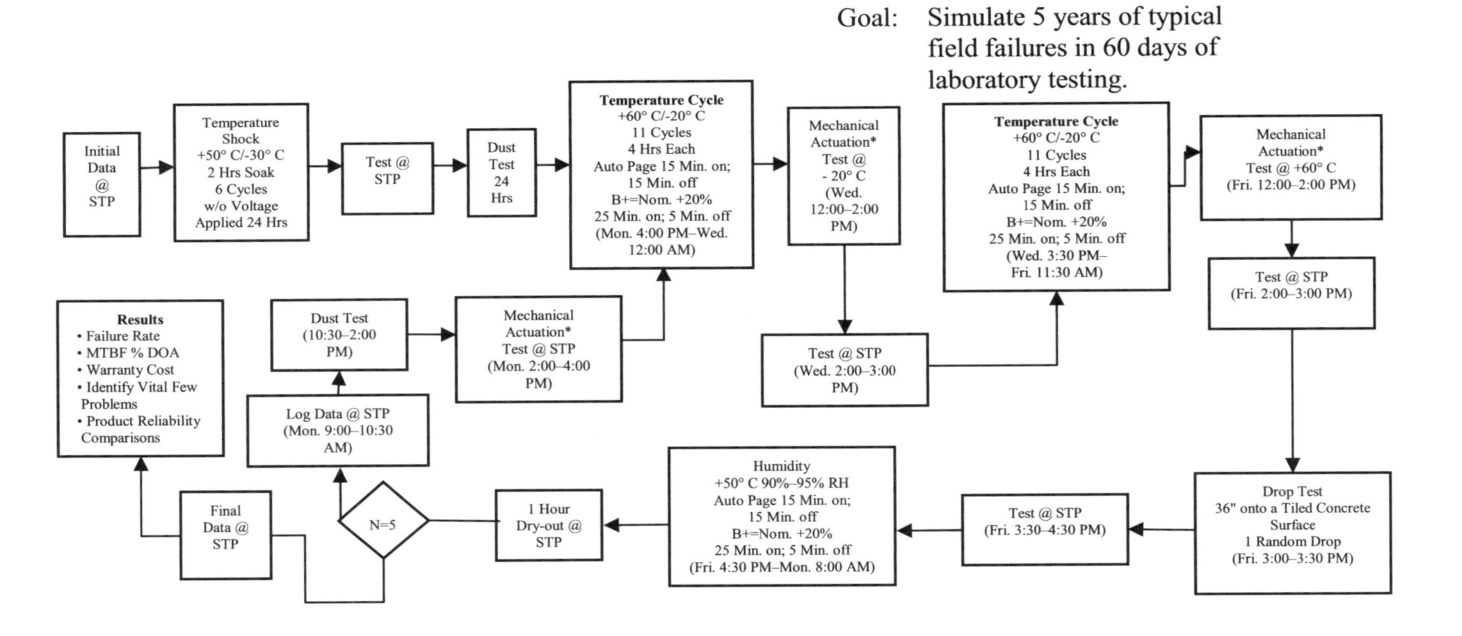

Acceleration factor = 43:1; 2 months alt = 5 typical field years.
*Exercise the switches, volume potentiometers, and mechanical moving parts 15 times.

generated, all right. However, there were several shortcomings in these techniques:

1. The demonstration tests took too long—months versus hours for MEOST.
2. Repeated tests could not generate the same failures.
3. Each new product had to go through different and repeated sequential tests in order to find the right stresses and their sequence.
4. Important failures caused by the interaction effects of stresses simultaneously applied were missed altogether.
5. Any design improvements had to be validated by another two months of testing.

Yet, this type of testing, one environment at a time, is still prevalent in the automotive industry and in the specifications of the Society for Automotive Engineers (SAE).

Having stressed the shortcomings of the 1) mathematical approach, 2) reliability prediction, 3) reliability estimation, and 4) reliability demonstration, we can now examine a halfway station—HALT/HASS—on our long journey to MEOST.

CHAPTER 7

Highly Accelerated Life Tests—HALT and HASS

"The acronym HALT is appropriate.
Having climbed up the reliability mountain,
It dawdles at Base Camp No. 1."

—AN INTERNAL COMPANY MEMO

HALT Background

Developed by Dr. Gregg K. Hobbs in the 1980s, HALT (Highly Accelerated Life Tests) has gained considerable currency in advancing the frontiers of reliability demonstration. It has been used in aerospace work and among some electronics companies. It has succeeded in improving reliability by factors of 5:1 in short design validation tests. Its principles are somewhat similar to Multiple Environment Over Stress Tests (MEOST), where they will be discussed in greater depth in Chapter 10. This chapter briefly outlines HALT methodology, its advantages and disadvantages, as well as those of HASS

(Highly Accelerated Stress Screening), which is a companion piece to HALT.

HALT Methodology

Performed at the prototype stage of design, HALT stresses a product well beyond design specifications right up to destruct levels, or the fundamental limit of technology. One interpretation of a destruct level is that it's a stress level that, when reduced, fails to make a product recover. Another interpretation defines destruct stress at that level where a small (further) level of stress causes large increases in the number of failures. A third interpretation is the fundamental limit of technology (FLT), which is defined as that level of stress where the product disintegrates. As an example, plastic softens to a point of no return at around 100°C, which then becomes its fundamental limit of technology. A similar level for solder, used in printed circuits, would be its melting point temperature of 250°C.

HALT generates multiple failures—a few in the lower stress levels beyond design limits, and a large number of rapid failures as the stresses approach destruct levels. It specifies that every single failure be analyzed and corrected. It calls for as large a sample size as possible, although the quantity would, of necessity, be limited at the prototype stage of design. It starts with application of one stress at a time, then recommends more combined stresses, which are usually confined to thermal cycling and vibration. Finally, HALT calls for repeated tests until *all* failure modes are corrected.

Table 7-1 lists the advantages and distinct disadvantages of HALT as a reliability demonstration technique.

Table 7-1. Advantages and disadvantages of HALT.

Advantages	*Disadvantages*
• Good track record in aerospace and electronics industries. • Better reliability improvement than traditional methods. • Forces failures in hours and days vs. months and years of field life. • Reduces cost of field failures. • Can be used at all levels of production—system, subsystem, and component. • Commercial test equipment available.	• Less successful for mechanical products, especially large sizes. • Reliability improvements not as spectacular as claimed. • Extending each stress test to destruct levels or the fundamental limit of technology (FLT) creates artificial failures that may never be replicated in the field. • Becomes cost- and time-prohibitive to analyze every stress failure down to root cause and correction. • Primary reliance is on each stress by itself; then on just two stresses—thermal cycling and vibration. • Large number of units recommended for tests. • HALT cycle repeated for each design change. • Units not returned from the field for further failure probing. • Not capable of reliability prediction.

Some Claimed Successes for HALT

- A computer peripheral device registered a 4.7 times increase in mean time between failures (MTBF).
- A power supply that had a "plug and play" rate of 94 percent (i.e., a failure rate of 6 percent) increased to a rate of 99.6 percent after HALT (i.e., a 0.4 percent failure rate). That may sound very encouraging, but a failure rate of 0.4 percent—or 4,000 parts per million

in production—is still enormously high when compared with failure rates of 20 parts per million after one year in the field with MEOST.

Highly Accelerated Stress Screening (HASS)

HASS is a sequel, in production, to HALT in design. Following is a brief synopsis of its definition, objectives, benefits, and methodology. (The HASS methodology is similar to HALT, but it stresses only to lower operational levels. Those levels are covered in Chapter 10.) Table 7-2 lists the pros and cons of Highly Accelerated Stress Screening.

Prerequisite	Full-fledged HALT in design
Definition	A 100 percent test screen with stresses higher than field stress—high enough to catch potential field defects but leave the rest of the product with over 80 percent of its useful life

Objectives

- ❑ Convert latent failures to patent failures with minimum cost and time.
- ❑ Reduce infant mortality and warranty costs.
- ❑ Reduce workmanship and supplier defects in production.

Benefits (Claimed)

- ❑ Increased reliability—more robust products
- ❑ Increased out-of-the-box quality

- ❑ Reduced manufacturing test time
- ❑ Faster time to market (a disputed benefit)

Methodology

- ❑ Select appropriate stresses (usually thermal cycling and vibration).
- ❑ Determine operational levels (OL) for stress (i.e., balance between finding latent defects and not using up more than 10–20 percent of useful product life).
- ❑ Use a sample size of 100 percent of production units.
- ❑ Step-stress, as in HALT, then analyze *each* failure and take corrective action.

Proof of Screen (to Verify HASS Effectiveness)

- ❑ Repeat complete HASS cycle (as outlined above) ten times.
- ❑ If failures occur, lower OL and rerun; if there are no failures, HASS OL levels are acceptable.
- ❑ Deliberately induce seeded samples to test HASS's ability to catch them.

HASS Optimization

- ❑ Run four HASS cycles on 100 percent of production units.
- ❑ If failures are not detected in first cycle, increase OL.
- ❑ Repeat proof of screen.
- ❑ Rerun the four HASS cycles until all failures occur in the first HASS cycle.
- ❑ If not successful, go back to HALT and start all over.

Table 7-2. Pros and cons of HASS.

Pros	*Cons*
• Proof of screen is a good validation of HASS principles. • Uses "seeded samples" (i.e., introducing deliberate defects to test HASS power of screening). • Screen time is quick—both for vibrations and thermal cycling.	• 100% HASS tests in production is a brute-force approach to reliability, almost as bad as burn-in. • There is no guarantee that all defects can be detected despite "proof of screen." • There is not much confidence that useful product life has not degraded by using HASS. • Only one or two stresses are employed—thermal cycling and vibration. • Proof of screen, seeded samples, and optimization are long and cumbersome procedures that can add costs and increase production cycle time. • No reliability prediction is attempted or possible.

Other Environmental Stress Screening (ESS)

ESS came into vogue twenty-five years ago as an alternative to traditional reliability demonstrations such as burn-in and Milspec series. HALT and HASS were the offsprings of ESS. Other systems are:

- ❑ Reliability Environment Test (RET)
- ❑ Accelerated Reliability Test (ART)
- ❑ Accelerated Stress Test (AST)
- ❑ Stress for Life (STRIFE)
- ❑ Failure Mode Verification Testing (FMVT)

All of these methods are similar to HALT, with only minor differences in reliability philosophy, practices, and interpretations. FMVT comes closest to MEOST, but because it continues until rapid failures occur, it prolongs test time and adds to test costs with an analysis of even artificial failures.

Highly Accelerated Stress Audit (HASA)[19]

One of the most important weaknesses of HASS as a production stress screen is that it is performed on 100 percent of the product in production, consuming labor and equipment (chamber) cost and—most crucial of all—valuable cycle time. Furthermore, it consumes at least 10 percent of product life and can weaken the reliability of the product that it is trying to assess.

HASA overcomes the weakness of the HASS 100 percent test in production by kicking in a simple plan (an audit) if the risk of failure is minimized. It starts with a 100 percent test, just like HASS, but it allows a smaller sample size if the total number of failures in HASS testing falls below five. HASA is used where the production quantities would be so large as to render 100 percent stress testing prohibitive in terms of costs and cycle time.

There are two primary weaknesses in applying HASA, however:

1. It allows the accumulation of five failures before sampling can replace 100 percent stress tests. This is contrary to reliability objectives to permit even two failures.
2. Based on β risks (i.e., the risk of accepting a product when it should be rejected), the sample size itself is enormous, as are the number of days of test and the number of failures.

As an example, for an allowable defect rate of 2 percent, the sample size (instead of 100 percent) would be 2,223 units, tested for twenty-four days and allowing 120 defects—an enormous waste of time and money.

Conclusion

HALT and HASS are base camps in our reliability ascent to the pinnacle of MEOST. Their greatest contribution is in alerting a skeptical and conservative reliability profession to the benefits of overstress testing.

Part III will explore the infrastructure needed in design for MEOST, the comprehensive preparation for MEOST, and the remarkable eight stages of MEOST.

PART III

The Climb to MEOST—The Mount Everest of Reliability

CHAPTER 8

Base Camp 1—Design Reliability Infrastructure

"Over 80 percent of new products end up on the ash heap of the marketplace."

—Bob King, *Better Design in Half the Time*[20]

Design: The Bête Noire of a Company's Failures

Product quality, reliability, cost, and market launch times are dominated by the design function.

- ❑ Over 80 percent of poor quality is caused by design.
- ❑ Over 90 percent of field failures are the result of poor design.
- ❑ Most product recalls from the field have their origin in design.
- ❑ Most lawsuits can be traced to improper design.

- ❑ Some 70–75 percent of product costs are a function of design.
- ❑ Long design cycle times give the keys to competition.
- ❑ Creativity and innovation are the victims of design bureaucracy.

If the objective, therefore, is to abolish field failures, the starting point has to be design and the creation of a supporting infrastructure. That infrastructure consists of:

1. Organization of the design function
2. Management guidelines
3. Techniques to capture "the voice of the customer"
4. Quality/reliability cost and cycle time in design
5. Creativity stimulation (for releasing the engineer's bottled-up, creative genie)

1. Organization of the Design Function

Traditionally in product design, the engineering function operates in isolation with a "we know what is good for the customer more than the customer does himself" attitude. The result is a half-baked product that is tossed over the wall (like a ticking time bomb) for the poor production department to catch and, eventually, for the customer to throw out the window.

(1) Concurrent Engineering

In the last fifteen to twenty years, companies have embraced the principle of a team approach—called concurrent engineering or simultaneous engineering. The first step in this principle is to add manufacturing expertise to the design team to facilitate design for manufacturability (DFM). The second step is to add the disciplines of quality, service, purchasing, and finance at the start of design. The third step is to add partnership suppliers to the team, which is led by a project manager or program manager. Unfortunately, even though concurrent engineering has replaced "closet engineering" in progressive companies, the team's effective functioning has been hampered by:

- Part-time effort, with the team members being pulled away by their functional managers for other chores. Part-time effort leads to part-time results.
- The team leader not being given absolute responsibility and authority to move the project forward. (In Japan, the team leader is virtually a czar—from project inception to first delivery in the field.)

(2) The Contract Book

In the past, there has been excessive finger-pointing between management and the design team about responsibility for failures in product launch. Management has blamed the team for not meeting stated targets for quality, cost, and design time. The team, in turn, has blamed management for poor resource planning, funding, and manpower. (This finger-pointing with crossed arms is often called the company salute—blame somebody else for your sins.)

One innovative remedy for this situation is the use of a contract book—a written agreement where management commits to provide the necessary manpower and capital, while the team pledges to meet the agreed-upon quality/reliability, cost, and design-cycle time goals.

(3) The Milestone Chart

Table 8-1 charts a twenty-step progression of a design from the concept stage up to production (job 1). It shows:

- The team member with prime responsibility (P) for each step, as well as those team members with contributing responsibility (C) for that step.
- Sign-off authority (S), which is generally executed by senior management at critical stages or steps in the design's progress, either to allow the project to continue or to terminate it if the quality, cost, and cycle time of the project are falling behind targets and recovery is deemed difficult.
- Sign-off authority (S) by the program manager (or team leader) during three important design reviews and at the start of production, to modify the scope or direction of the project.
- The major tools that should be employed at each step of the design. This is most important because without these powerful tools the design is likely to be stillborn. Of special interest are the twenty-first century tools at the prototype stage of design—Design of Experiments, Multiple Environmental Over Stress Tests, value engineering, group technology, and design for manufacturability.

2. Management Guidelines

There are several management guidelines that should govern a design. They can be considered part of policy deployment and incorporated into the contract book.

Table 8-1. New product introduction: milestones and responsibilities chart.

No.	Milestone	Sr. Mgmt	Pgrm Mgmt	Design Ldr	Mfg. Ldr	Quality	Sales / Mktg.	Service	Sourcing	Finance	Major Tools
A.	**Organization**										
1.	Program Team Kickoff	S	P	C	C	C	C	C	C	C	Concurrent Engineering, Contract Book
B.	**Mgmt. Guidelines**										
2.	Max. 25% Redesign	S	P	P		C	C			C	Policy Deployment
3.	Stream of Rapid New Products	S	P	P		C	C			C	Lessons Learned Log, Cycle Time Reduction
C.	**Voice of Customer**										
4.	Elements of Customer "WOW"		P	P	C	C	C	C	C	C	"Bhote's Law," Delight Features
5.	Customer Specifications		P	P	C	C	C	C	C	C	QFD, Mass Customization, Scatter Plots
D.	**Design Quality/Reliability**										
1.	Feasibility Study										CAD, Software Architecture
2.	Preliminary Design			P		C		C	C		Derating, FMEA, FTA, PLA
3.	First Design Review		S	P	C	C	C	C	C	C	Checklist
Prototype Design			C	P	C	C	C		C	C	DOE, MEOST, Iso-Plots
Quality Systems Audit			C	C	C	P	C	C	C		The Ultimate Six Sigma Assessment
Second Design Review			S	P	C	C	C	C	C	C	Checklist

(continues)

Table 8-1. (continued)

Pilot Run		C	P	C	C		C	C	C	B vs C, Positrol, Process Certification, Precontrol
Design for Manufacturability			P	P	C					Boothroyd-Dewhurst Scoring
Design Cost Reduction										
Total Value Engineering		C	P	C	C			C	C	VE Job Plan, “Fast” Diagram
Part Number Reduction			P	C	C	C		C	C	SBU, Product, Model Reduction
Early Supplier Involvement		C	P		C		C	P	C	Cost Targeting, Financial Incentives
Patent Study		C	P						C	
Design Cycle Time Reduction										
Outsourcing		C	P	C	C			P	C	“Black Box” Supplier Design
“Human Inventory”		C	C					C	P	Integration of Cost-Time Curve
Third Design Review		S	P	C	C	C	C	C	C	Checklist
Product Launch										
Customer/Field Test		C	P	C	P		C			DOE at Customer Site
Management Review	S	P	C	C	C	C	C	C	C	Authorization for Full Production
Production Run		S	C	P	C		C	C		Mini-MEOST, TDPU, COPQ
Management Audit	P	P	C	C	C	C	C	C	C	Lessons Learned
Initial Customer Feedback		P	C	C	P	P	C	C	C	Survey Instrument

Responsibility Codes:

S = Sign-Off Authority
P = Prime Responsibility
C = Contributing Responsibility

(1) Maximum 25 Percent Redesign Rule

This means that no more than 25 percent of the old design (by parts volume) should be changed and incorporated into the new design (unless there is a major platform change, which may occur only once in a few years). In the West, there is an unseemly urge for a 100 percent redesign to boost an engineer's ego and to see his name etched into the product in perpetuity. This results in long delays in launching the product, with competition forging ahead at the marketplace.

(2) A Stream of New Products Rapidly and Frequently Introduced

Another policy deployment that the maximum 25 percent redesign rule achieves is that it allows a company to introduce a new product every few months, instead of every few years. This facilitates new features, new innovations, and rapid adjustments to changing customer needs. This constant stream of new products moves an organization four to six generations ahead of the competition, which is left in the dust of the whirlwind of introductions. The classic example of this type of product deployment is NIKE, Inc., a company that introduces small changes, but so frequently that it keeps its competitors off-balance and frustrated.

(3) Supply Management Directives

Early in the design cycle, management must stipulate to the team a few key practices in its dealings with suppliers. Among them:

- ❑ Selecting key suppliers based on ethics, trust, and mutual help to promote a win-win partnership.

- ❑ Never discarding a partnership supplier, particularly one of long-standing, for the lure of a supposedly lower price from another supplier, which may result in higher ownership costs.
- ❑ Never discarding a partnership supplier because of electronic procurement and bids to the whole world to find the lowest-cost supplier. This is, to put it bluntly, dumb. Unfortunately, the automotive companies and other reputable corporations are becoming suckers for the siren song of electronic procurement.
- ❑ Not squeezing a supplier for a price reduction on the economic strength of a "Big Brother" customer. The supplier may comply in order to retain the business, but will attempt to recover his "fair" price through price adjustments for changes in specifications, quality shortcuts, and schedule slippages. When this happens, the principle of a win-win partnership is broken and, like Humpty-Dumpty, it can never be put back together again.
- ❑ Establishing a policy with partnership suppliers for continual quality, cost, and lead-time improvements, but allowing and coaching the supplier to secure higher profits. In short, this establishes a ceiling for supplier's prices and a floor for supplier profits.
- ❑ Outsourcing design to qualified suppliers on those parts/products where the company does not possess great competence. This is one of the best ways to save design time and design costs.
- ❑ Outsourcing not just piece parts, but higher-level assemblies and "black box" designs that are not the company's core competencies.
- ❑ Encouraging the team's acceptance of early supplier involvement (ESI) at the prototype stage of design as one of the best ways to reduce costs and improve quality.

- ❑ Using cost targeting. This means that the company—not the supplier—prescribes a price for the part/product, based on calculated material and labor costs, rather than the old three-quote syndrome.
- ❑ Establishing a competent commodity team to actively and constantly help and coach the partnership suppliers to reduce their high defects, cost, and cycle time, and then sharing equitably in the savings.
- ❑ Establishing reliability as a specification and designing financial incentives and penalties for attaining and not attaining, respectively, these reliability targets.

(4) A Reliability Philosophy on Parts

- ❑ Ninety percent of the parts must have a proven track record of reliability. (There is undoubtedly a need for using state-of-the-art new parts, but such experimentation should be kept to a minimum. MEOST success is a proven way to introduce new parts.)
- ❑ All critical parts must be derated (i.e., have a factor of safety) in order to enhance part reliability.

(5) Lessons Learned Log

There has been a predilection among most companies to sweep design mistakes under the rug. The objective of a lessons learned log—either a paper or a computer record—is not to embarrass the design team or punish it, but to ensure that the sad history of accumulated mistakes is not repeated, either during the next product design or by the next generation of engineers.

(6) Reverse Engineering

Known also as competitive analysis, reverse engineering should be standard practice in any company's design cycle. It works as follows:

- The competitor's product is stripped down and evaluated part by part for the design, materials, assembly, and cost (through cost targeting).
- MEOST is performed to compare failure modes, stress-to-failure, and time-to-failure of the company design versus the competitive product.
- Quality function deployment studies are conducted, wherein the customer rates the company against the best competition for each requirement.
- Benchmarking is employed to determine best design practices in other companies.

3. Techniques to Capture the Voice of the Customer

In the past, designs have been determined by "the voice of the engineer" or "the voice of management," both of whom think that they know more about what the customer needs and requires than the customer himself. This ostrich-like, head-in-the-sand approach results in new designs that end up on the ash heap of the marketplace.

(1) Quality Function Deployment

In the last twenty years, quality function deployment (QFD)[20] has been used to capture the voice of the customer and translate it into engineering specifications. One of its limitations, however, is that QFD concentrates on the technical performance of a product, rather than a holistic view of customer requirements.

(2) Mass Customization

A more recent technique developed in the last fifteen years is based on the premise that each customer's requirements are unique, and they are different from any other customer. The proliferation of options that mass customization entails—sometimes as many as one million—makes it a difficult task when it comes to production. But with a few tips, mass customization can work very well.

Tips for Making Mass Customization Productive

1. Reduce the number of customers, especially the 20–30 percent of customers who usually account for 20–40 percent of a company's profit *loss*.
2. Reduce the number of proliferating options. Typically, the number of permutations of options can rise to 500,000 to 1,000,000. At least half such options are unprofitable, and they should be jettisoned.
3. Postpone product differentiation (customization) to the last step in the customer chain. If possible, customization should be done, in descending order, 1) at customer site (preferred), 2) at the dealer, 3) in packing, and 4) in final assembly and test.
4. Modularize the design.
5. Maximize standardization or, at least, commonality.
6. Produce modules in parallel, not in sequence (i.e., in series).

7. Introduce a pull (kanban) system with key suppliers to prevent stock-outs.
8. Program instant changes with programmable integrated circuits, such as PROM and E-PROM.
9. Reduce cycle time to the customer.
10. Increase inventory turns.

(3) Elements of Customer Enthusiasm

A more practical starting point is a network of the elements that combine to create customer enthusiasm and value. Figure 8-1 depicts such a network of twenty elements. But which element is most important to a customer? There is no easy answer. No single element is most important to all customers at all times, in all places. However, two principles are vital. The first, called "Bhote's Law" by my students in graduate school, states: It is that element missing from your product, and which is important to your customer, that a company must pursue with laser-like intensity. The second principle states that the design should introduce features unanticipated by the customer, but whose creation results in a rush of customer delight. The Japanese capture this spirit of unexpected delight in going from "Atarimae Hinchitsu"—taken-for-granted quality—to "Mirya Kutcki Hinchitsu"—quality that fascinates, bewitches, and delights.

(4) Challenging Specifications—or Why 90 Percent of Them Are Wrong

Forty years of experience working with engineers leads to the firm conclusion that almost all specifications are wrong. They are either too loose or too tight, but invariably wrong. They are pulled out of the air. The practice is labeled "atmospheric

Figure 8-1. A network of elements of customer enthusiasm—for products.

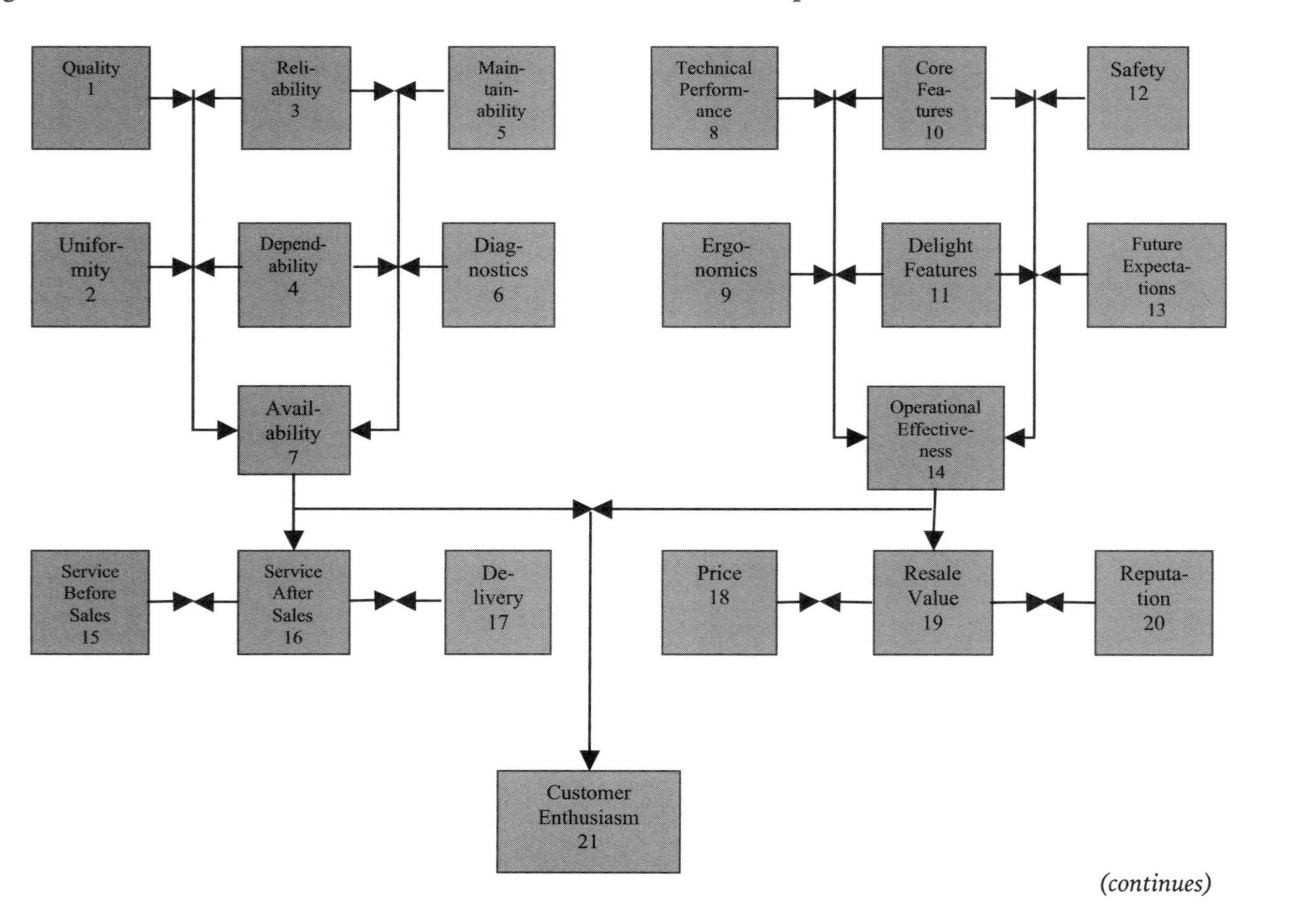

(continues)

Figure 8-1. (continued)

1. Quality: Toward Zero Defects
2. Uniformity: Toward Zero Variation
3. Reliability: Toward Zero Field Failures
4. Dependability: Toward Lifetime Guarantees
5. Maintainability: Toward Accurate, Fast, Low-Cost Repair
6. Diagnostics: Toward Customer Self-Diagnostics
7. Availability: Toward 100% Uptime
8. Technical Performance: State-of-the-Art Technology
9. Ergonomics: Styling, Color, Ease of Operation–"User Friendly"
10. Core Features: Expected by Customer
11. Delight Features: Unexpected Features That Thrill Customer*
12. Safety: Of Product and to User Product Liability Prevention
13. Future Expectations: Anticipating Needs
14. Operational Effectiveness: Integration of Boxes 8 through 13
15. Service Before Sales: Sales, Cooperativeness, Communication
16. Service After Sales: Sustained Contact and Interest After Sales
17. Delivery: Short Cycle Time
18. Price: Cost Below Competition
19. Resale Value: High Percent of Purchase Price
20. Reputation: Image, Perceived Quality
21. **Customer Enthusiasm: Value, Delight, Loyalty**

---

*From *Atarimae Hinchitsu* to *Mirya Kutcki Hinchitsu*: "Taken-for-granted quality to quality that fascinates, bewitches, delights."

analysis." Why are specifications so vague and arbitrary? There are several reasons:

- Specifications are lifted, in total, from older designs and drawings.
- Engineers rely on boilerplate requirements or supplier-published specifications.
- Computer-aided tolerances are accurate only if the formula governing the relationship between the output (i.e., the dependent variable) and the independent variables is known, which is rarely the case in complex designs.
- Worst-case analysis is performed. Because of the very low probability of such occurrences, there is an appreciable addition to cost with little added value.
- Statistical tolerancing—a formulaic approach—is too generalized to be applicable in all situations.
- Engineering conservatism (otherwise known as "covering your hide") is a prevailing attitude. Engineers know that tight specifications and tight tolerances may add to product cost. But loose specifications and loose tolerances, resulting in product failures, may cost them their jobs.
- Reliability is not a specification. Even in the rare case that it is, reliability is not tested or demonstrated. As a result, most reliability estimates, such as mean time between failures, border on fiction.
- Field environments/stresses are not measured, combined, or simulated in the design laboratory.
- Ergonomics are not considered sufficiently. Engineers do not take into account the customer's lack of familiarity with the product, or its misuse by the customer.

- ❑ Product safety is not pursued aggressively.
- ❑ Product liability prevention is not practiced sufficiently. It should be a design "must."
- ❑ Built-in diagnostics are not yet a design culture.
- ❑ Correlation studies, which chart levels of customer preferences and perceptions against levels of one or more parameters, are not conducted.

Specifications Start with the Customer

This last point is especially important. Only the customer can define his requirements. Sometimes, these requirements may be nonquantitative and subjective. Yet, when translating these customer requirements into engineering specifications, the design team must determine the correlation of the customer requirement (even if it is on a subjective scale of, say, one to ten) on the Y-axis (dependent variable) with the associated engineering parameter on the X-axis (independent variable) using the scatter plot technique.[3]

A strong correlation (narrow parallelogram with tilt) confirms the validity of the engineering parameter. A weak correlation (wide parallelogram with little tilt) indicates that the engineering parameter is unimportant and its tolerance can be opened up to the most economic levels.

CASE STUDY: EDGE DEFECTS IN CONTACT LENSES

A manufacturer of contact lenses was convinced that cosmetic edge defects in its product had to be rooted out as objectionable to the end-user—the lens wearer. These edge defects included scratches, chips, and inclusions around the periphery of the lens. It was difficult to see the edge defects with the naked eye. High-powered microscopes had to be used. Each lens had to go through four or five inspection stations to brute-force its way to quality and customer requirements.

The company spent millions of dollars on this sorting process. Then a new competitor arose on the horizon and captured a large market share at the expense of the established company. Certain that the competitor had solved the problem of edge defects, the manufacturer did a reverse engineering study and found—to its utter amazement—that the competitor's lenses had far more edge defects than its own lenses.

A marketing study, using scatter plots, was instituted to determine the correlation between edge defects (independent variables) and the customer's perception of satisfaction on a scale of one to ten (the dependent variable). The study revealed that there was little correlation between edge defects and customer satisfaction. Customers did not care about the edge defects because they were only on the periphery of the lens and could not be seen by the naked eye. Most important, the defects did not affect vision, which is concentrated in the center of the lens.

A further study found that the competitor's product was superior because of its 1) closer adherence to the optical power specification of the lens and 2) greater wearing comfort. The manufacturer de-emphasized its brute-force correction of edge defects and concentrated on prescription accuracy with a C_{pk} of 2.0 and on wearing comfort; and it was able to restore its market share in eighteen months.

In some cases, the end-user—the real customer—is separated from the manufacturer (i.e., supplier) by an in-between customer, such as an original equipment manufacturer (OEM), who may be arbitrary and unreasonable about specifications handed down to the supplier. This dilemma can be overcome in two ways:

1. By involving the end user, OEM, and supplier in a joint correlation study.
2. By doing a sensitivity analysis, negotiated with the customer, using the relationship between incremental specifications and incremental costs. Sensitivity analysis is effective because it matches specification adjustments to the customer's pocketbook. For example, as shown in Figure 8-2(A):
    - If the customer sees great gains in a desired output for small cost increases (e.g., much better solderability for a slightly more expensive flux), he would tend to accept the cost increase.
    - If the customer observed minimal gains in a desired output—a reduction in the height of a tape deck, for example—for a large cost increase, he would reject the trade-off.

Similarly, in Figure 8-2(B):

- If the customer sees an appreciable cost reduction for a small reduction in a specification (e.g., reduced or zero burn-in time), he would accept the specification reduction.
- If the customer sees only a marginal cost reduction in return for a reduced specification (e.g., bond strength much reduced for a bond wire diameter reduction), he would veto the specification reduction.

Figure 8-2. Sensitivity analysis: incremental specifications vs. incremental cost.

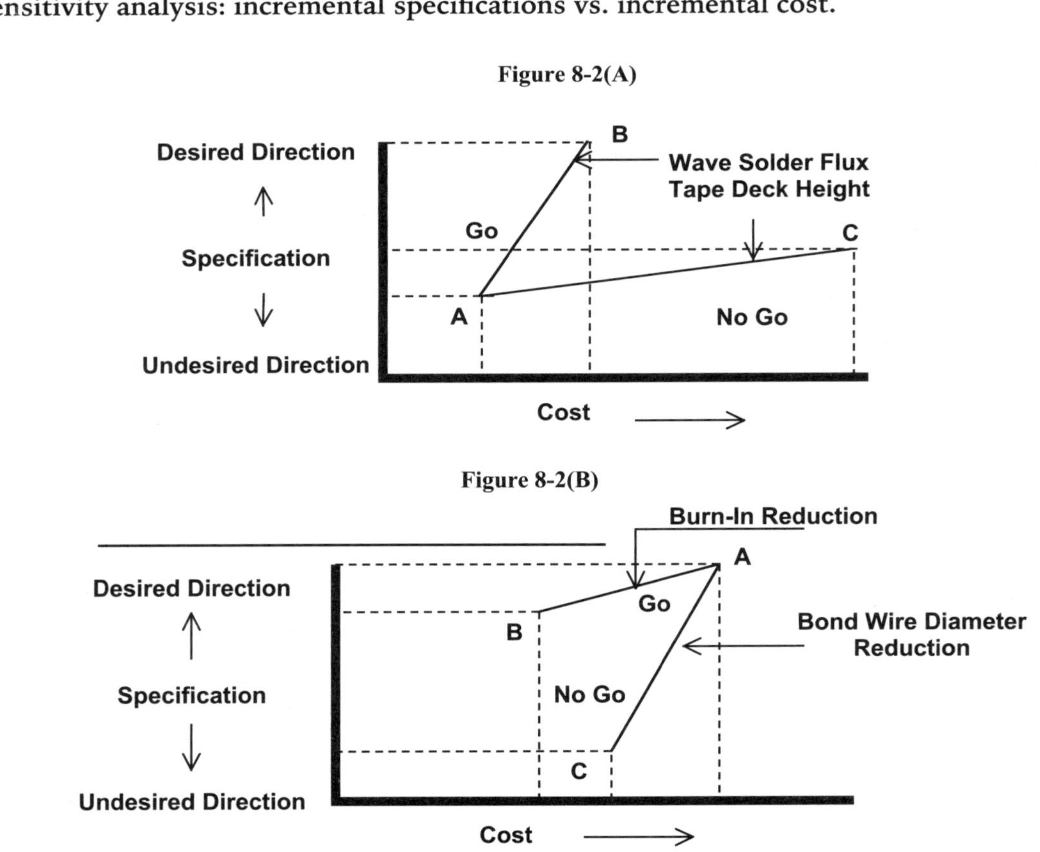

A Specification Checklist

Figure 8-1 depicted a network of twenty elements that together added up to customer enthusiasm. Most of these elements can be quantified with meaningful specifications, which the design team must determine. Table 8-2 shows a comprehensive specifications checklist to guide the design team in the formulation of such specifications.

4. Quality, Cost, and Cycle Time in Design

Even though reliability is the focus of this book, there are disciplines of quality, cost, and cycle time that are necessary ingredients of the design infrastructure and that impact reliability. These disciplines are explained in detail in *The Power of Ultimate Six Sigma.*[1] Only a few tools are highlighted here.

Quality Disciplines

- ❑ Design of Experiments (DOE)—Shainin/Bhote, not Classical or Taguchi DOE
- ❑ Process/product characterization and optimization
- ❑ Process certification
- ❑ Positrol
- ❑ Poka-yoke

Cost Disciplines

- ❑ Total Value Engineering
- ❑ SBU, product, model, and part number reduction
- ❑ Group technology (GT)
- ❑ Early supplier involvement (ESI)

Table 8-2. A specification checklist.

Customer Element	
Quality	• Are acceptance/rejection criteria established, and are they based on the customer's expectations, not the company's? • Are target COPQs, yields/cycle time, and TDPU (total defects per unit) established? • Are the Shainin/Bhote DOE requirements to prevent quality problems in production sufficiently well known and practiced?

Reliability	• Do targets (percent failure rate/year, MTBF, etc.) reach out to meet future reliability expectations, such as:
	  • Sharp reductions in the gap between failure rates of complex versus simple products?
	  • Component failure rates below one PPM/year?
	  • Use of incentive/penalty clauses for exceeding/not achieving reliability targets?
	• Are maximum field environments/stresses measured?
	• Are adequate test chambers in place to combine and accelerate stresses rapidly to reliably predict and prevent failure rates using MEOST?
	• Is there a firm derating (factor of safety) protocol in place, and is it followed?

Maintainability	• Are targets for mean time to diagnose (MTTD) and mean time to repair (MTTR) established?
Diagnostics	• Are MTTD targets helped by built-in diagnostics and MTTR targets by modular designs?
Uniformity	• Are target values and minimum C_{pk}s determined for key parameters?
Dependability	• Are warranty targets/times being extended toward lifetime warranties?
Availability	• Is there a targeted uptime as a percent of the total product use time?
Technical Performance	• Are target performance parameters established with the customer as the "supreme court," and are they correlated with customer expectations? • How do target performance parameters compare with the competition?
Safety Place	• Is a formal, written product liability prevention process in place?
Ergonomics	• Is user-friendliness tested with employee/focus groups, and how does it compare with the competition?
Future Expectations	• Are future expectations continually solicited from customers through mass customization?

(continues)

Table 8-2. (continued)

Customer Element	
Service Before Sales	• Are the sales/service parameters formulated to measure the effectiveness of service to the customer?
Service After Sales	• Are there targets for repair service accuracy and timeliness? • Is there a systematic and continuous follow-up of customers after sale?
Price/Cost	• Have target prices, targeted costs, and targeted product life been determined? • Have price elasticities been determined? • Have incremental specifications versus incremental costs been tested with key customers?
Delivery	• Is there a targeted design cycle time that is at least half (or less) of a previous design inventory, made up of design manpower, expressed in terms of a negative cash flow equivalent of months of sales?
Delight Features	• Are there a targeted number of delight features, unexpected by the customer, to generate customer "wow"?

Cycle Time Disciplines

- ❑ Outsourcing all except core competencies
- ❑ "Black box" vs. piece-part procurement
- ❑ Early supplier involvement (ESI)
- ❑ Mass customization
- ❑ Parallel vs. serial development
- ❑ Daily team interchanges
- ❑ Part number reduction, preferred parts, and parts qualification by partnership suppliers
- ❑ Design of Experiments (DOE)
- ❑ Multiple Environment Over Stress Tests (MEOST)

Design cycle time is crucial for corporate success. Short development cycle times result in 72 percent of revenues from first-to-market products; and a McKinsey survey found, among eighty automotive suppliers, that corporate success was a function of development time, not cost.

5. Creativity Stimulation: Releasing the Engineer's Bottled-Up, Creative Genie[1]

Creativity is a mental act that frees a person or organization from self-imposed boundries of common knowledge. It may also be defined as:

- Breaking out of traditional thinking
- An expression of the talent, heart, and essence of individuals and groups
- Deferred judgment

Creativity does not come naturally to all people. Yet its disciplines can be learned by any person. While creativity can be established in all areas of a company, it is best focused in the design function, where its needs and benefits are the greatest.

A Seven-Step Approach for Creativity Stimulation

1. *Orientation*. Determine the true nature of the problem or task, eliminate misconceptions, and determine related factors.
2. *Preparation*. Gather a storehouse of facts, general information, and past experiences.
3. *Analysis*. Consider the following questions:
    - Why does the problem exist?
    - What are the consequences of the problem and its cost?

- ❑ Are the items causing the problem really necessary? If not, consider eliminating them.
- ❑ Would a substitute item be better than a redesign?
- ❑ Is there a relationship between the subproblems?
- ❑ Has elapsed time changed the conditions?
- ❑ Is the problem really worth solving?

4. *Hypothesis*. Discard preconceptions and emphasize unlimited imagination (thinking outside the box). Neither critical evaluation nor negative thinking are allowed in this step.
5. *Incubation*. Allow time for subconscious thinking.
6. *Synthesis*. Put all the pieces back together after the analysis step.
7. *Verification*. Test and follow up to determine the effectiveness of the final solution. Do a B vs. C test (explained in Chapter 11).

A Blueprint to Raise Human Intelligence, Creativity, and Innovation for All

Training, important though it is, isn't enough. There must be leadership—not just at the CEO levels, but at all levels where people lead and influence others—and a nurturing, nourishing organization structure to maximize the potential for creativity and innovation. The elements of such an environment are:

- ❑ *Trust*. Bob Galvin, the retired chairman of Motorola, has stated: "One's creativity depends on interaction with others—others one trusts, others who feel trusted. For one to be unfettered in risking interaction with another, the other must know the trust of openness, objectivity, and complementary creative spirit . . . [I]n order to trust, one must be trustable. Trust is a power. The power to trust and be trusted is an essential prerequisite quality to the optimum development and employment of a creative culture."

- *Help and Guidance.* In Japan, the CEO looks upon the care and feeding of young employees as his primary responsibility. The help that enables people to grow is not a rigid master-to-apprentice regimen. It is guidance with a loose rein. Mentoring programs, which companies such as Ford Motor have institutionalized, are a good example of smoothing the road to creativity.
- *Freedom to Explore and to Make Mistakes.* The true leader, recognizing that he is not God and does not have all the answers, gives his people the freedom to explore their own pathways, create their own solutions, and even make their own mistakes. Freedom, however, is not a substitute for anarchy. The leader has to lead, set the direction, establish goals, and monitor results. Then, having done that, the leader gets out of the way. The mark of an outstanding leader, according to Jack Welch, retired chairman of GE, is the three Ds—direct, delegate, and disappear.
- *Leader-Follower in an Iterative Role.* Leadership implies that others will follow. But is the leader a breed apart, or is he a better follower? A leader has many roles, including that of:

    - Observer of the work his associates perform
    - Sensor of attitude, feelings, and trends
    - Listener to ideas, suggestions, and complaints
    - Student of advisers, inside and outside the institution

A leader is also the product of experience, both his own experience and that of others, and he may mimic other leaders who have earned his respect.

- *Team Synergy.* Today, the basic building block of an organization is not the department but the team. Not only are teams more productive—they have more fun. The team culture transforms the organization from separate, independent islands into a symbiotic interdependence. An esprit de corps develops, as in the world of sports.
- *Problem Solving.* One of the most powerful tools in the creativity arsenal is problem solving "by talking to the parts." *World Class Quality* details the various Shainin/Bhote Design of Experiments techniques that can coax answers from parts and processes much better than engineering guesses, hunches, theories, and opinions.[3] And the joy of success in the eyes of the team members is awesome to behold.
- *Support, Encouragement, Celebration.* The psyche of employees can be fragile. Bossy attitudes, body language that demeans, and a tin ear to ideas can ruin motivation, creativity, and growth. So can harsh layoffs and gross unfairness in income distributions for tangible gains generated by employees. By contrast, if the leader displays warmth and humility, has listening skills, is more of a coach and a teacher, supports the efforts of workers, encourages and cheers them on, and celebrates their "wins," the potential for advancement in creativity and innovation in every employee is unbounded.

CHAPTER 9

Base Camp 2—Essential Prerequisites for MEOST

"Those who ignore the lessons of history are condemned to repeat it."

—Old Proverb

A climb to Mount Everest requires taking an adequate amount of equipment, supplies, and food for the arduous journey. Similarly, a MEOST undertaking needs essential prerequisites, without which the effort can end up as a landslide.

1. Derating: Oxygen for the Climb to MEOST

One of the main reasons for poor reliability in products is that designers receive directives from management to cut costs. Engineers translate this message to mean they must select cheaper parts that are stressed to the maximum of their limits, rather than more expensive parts that are subjected to much lower stresses and, therefore, have much greater reliability. This is the eternal battle between lower product costs and the costs of field failure.

Many of our clients tell us, initially, that the higher warranty costs of field failures on a part are lower than the higher costs of a part that is deliberately stressed below its published rating. This is called derating. But this is a false trade-off. What these clients do not recognize is that warranty costs are only the tip of the iceberg. The total costs of such field failures include:

- ❑ Warranty Costs
- ❑ Out-of-Warranty Costs
- ❑ Costs of Downtime to Customers
- ❑ Field Trips
- ❑ Field Problem Solving
- ❑ Retrofit Costs and After-Sale Service Costs
- ❑ Litigation Costs
    1. Out-of-Court Settlements
    2. Court Judgments/Penalties (especially punitive damages)
- ❑ Costs of Customer Defections
- ❑ Costs of Loss of Referrals
- ❑ Loss of Reputation (i.e., public perceptions)
- ❑ Loss of Resale Value

It is no wonder that the most important thrust of a cost-reduction effort should be to identify, codify, analyze, and drastically reduce field failures. Derating is an essential prerequisite. It further reduces the need for frequent redesigns when successive MEOST trials repeatedly and needlessly fail the same overrated part in a product.

One of the most successful derating applications is in civil engineering. For instance, when designing a bridge, calculations are based on a stress or load that is three to six times what is likely to occur in practice. That is a factor of safety of three to six. Unfortunately, those safety margins are rarely considered in electrical designs, where the practice is to use safety margins, or derating, of 10 percent or less (once again because manufacturers worship at the sacred shrine of the lowest product cost).

In mechanical designs, a factor of safety of 2:1, or a derating of 50 percent, is generally practiced. But even here, catastrophic slippages occur, as in the case of the Challenger spaceship disaster in 1986, caused by the lowly O-ring not being sufficiently derated for low temperature.

To ensure adequate safety/reliability margins, the parts should "loaf"—that is, they need to be stressed so low that they can last for a long, long time. By contrast, high stresses cause weak parts, which then fail prematurely. The basic guidelines are:

Derating for mechanical parts	50 percent
Derating for electrical parts	40 percent

Figure 9-1 shows how steeply a failure rate on a bipolar linear integrated circuit (IC) can rise, once it gets past the knee of the curve around 105° C. Derating should always be considered for temperature as well as electrical parameters such as current, voltage, and power.

CASE STUDY: THE 1 AMP DISASTER

The history of reliability disasters is replete with design myopia. A multinational company used a 1-amp diode in a diode trio subassembly that formed part of an alter-

Figure 9-1. Failure rate multiplier vs. temperature bipolar linear.

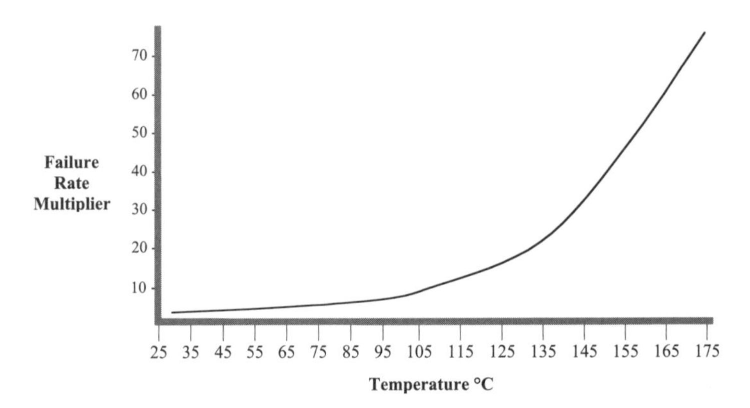

nator in a passenger car. The failure rates, low in early production, soon escalated into a major calamity, costing the company millions of dollars. The engineers had been reluctant to derate the diode because a higher-rated part would cost a nickel more! Finally, the threat of the total loss of the alternator business forced the company to design a 3-amp diode (i.e., a derating of 67 percent, or a factor of safety of three). The result: On 800,000 new alternators with the larger diode, the total failure was one—a failure rate of 1.25 parts per million (PPM), as compared to the previous failure rate of 15 percent or 150,000 PPM.

Workshop Exercise

A heavy truck manufacturer sells 50,000 trucks a year at a selling price of $60,000 per truck. It is experiencing an 8 percent failure rate per year on its alternator. The warranty period is one year. The alternator costs $100, but the dealer reimbursement cost is $150.

The engineering department calculated that the alternator was operating at 90 percent of the maximum rating allowed by its suppliers (i.e., a derating of 10 percent).

Because of the current high failure rate, it was considering a larger alternator with a derating of 60 percent (i.e., at 40 percent of maximum rating allowed). The projected failure rate was 50 parts per million per year. But the cost increase would be $15 (from $100 to $115).

Questions

1. Determine the total cost of the current alternator failure versus the total cost of the new alternator. Is the extra cost of the new alternator justified?
2. Assuming that only one percent of the dissatisfied 8 percent of current customers defect to competition, what would the extra cost impact be of the current alternator?
3. Assuming that five of these dissatisfied customers, who would each be likely to buy four such trucks in their working lifetimes, are so turned off that they would never buy this manufacturer's truck, what would be the extra cost impact of the present alternator?
4. Assuming that these five disgruntled customers each told ten of their friends and colleagues about their bad experience with the manufacturer, and that only one from each group of ten boycotted the company, estimate the extra cost impact of the current alternator.
5. Assuming that the manufacturer, to increase market share, offered the trucking customers a downtime allowance of $500, what would be the cost of the current alternator?
6. Assuming that the manufacturer would buy back 1,000 of the trucks from the customers at a cost of $1,000 above fair resale value (to make up for the poor reliability), what would be the loss to the manufacturer of the current alternator?

Answers

Figure 9-2 shows that the warranty cost is the tip of the iceberg when total costs are calculated.

Question 1. Comparison of Current vs. New Alternator

a. Current Alternator Cost = Failure Rate × Total Trucks × Replacement Cost, or 8% × 50,000 × $150 = $600,000.
b. New Alternator Cost = Extra Cost × Total Trucks, or $15 × 50,000 = $750,000.
   (The warranty cost of the new alternator would be negligible: Warranty Cost = Failure Rate × Total Trucks × Replacement Cost, or (50 × 10–6) × 50,000 × $150 = $37.5.)

Conclusion: The more reliable alternator would be *$150,000 higher* per year, despite the current alternator failure rate.

Question 2. Cost of Customer Defection

Number of Defecting Customers = 1% × 8% × 50,000 = 40
Total Cost of Defecting Customers = 40 × $60,000 = $2,400,000

Conclusion: The cost of even a tiny percentage of customer defection (0.08 percent) is four times the current warranty cost.

Question 3. Lifetime Costs of Five Permanently Defecting Customers

Lifetime Costs = 5 × 4 × $60,000 = $1,200,000

Conclusion: The cost of just five customer defections over their lifetime (possibly four lost purchases per person) is twice the warranty cost.

Question 4. Loss of Referrals (the Bad-Mouthing Factor)

With five referrals lost for the customers' lifetime, the loss is 5 × 4 × $60,000 = $1,200,000.

Conclusion: Even a tiny loss in referrals to potential customers adds up to a cost twice the warranty cost.

Question 5. Downtime Allowance to Unhappy Customers

With a total of 4,000 unhappy customers compensated for their downtime losses, at $500 each, then the total cost is 4,000 × $500 = $2,000,000.

Conclusion: In industries such as transportation, customers are far more concerned with downtime than with just warranty compensation. Here, the total costs are more than 300 percent of the warrant cost.

Figure 9-2. Workshop Exercise: Warranty is the tip of the iceberg.

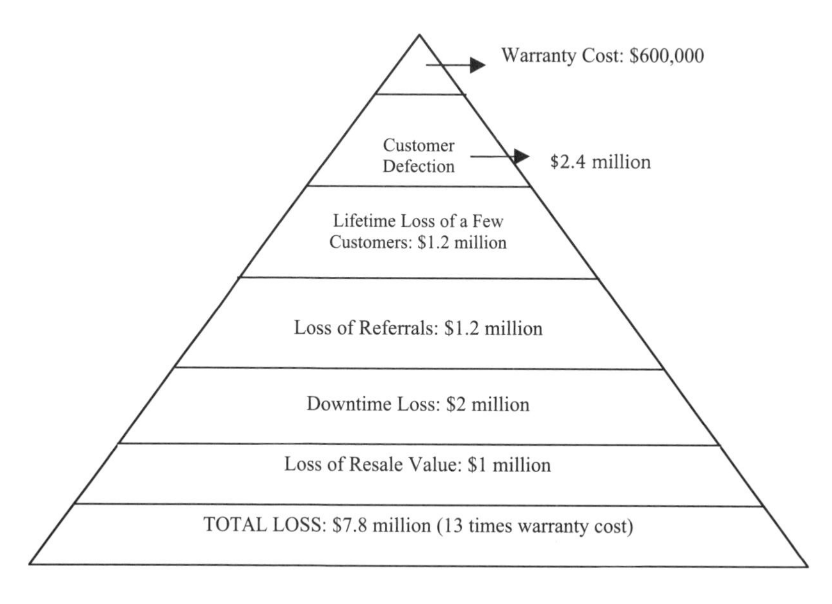

Question 6. Loss of Resale Value

With the manufacturer buying back trucks from 1,000 unhappy customers at a cost of $1,000 (above fair resale value), the total costs are 1,000 × $1,000 = $1,000,000.

Conclusion: A company pays for poor reliability by being forced to buy back its product above fair resale value.

2. Temperature: The Enemy of Reliability

A second design concern for reliability—besides insufficient derating—is temperature. High temperature accelerates the physics of failure in many parts (although low temperature can also accelerate brittle parts).

Thermal Scanning

Employing infrared scanners is a useful pre-MEOST technique to "sniff out" areas of a product with a high temperature rise. The design or layout can then be changed, or heat sinks used, to reduce these high temperatures. In a few special cases, it is not only important to measure the ambient temperature around such parts (especially in MEOST setups), but also to calculate temperatures inside the part, such as junction temperatures, which could be several degrees higher.

3. Modular Designs

Easily plugged-in and plugged-out modules have become popular in design development ever since Motorola introduced its famous "works in the drawer" television sets in the 1960s. There are several advantages to modular designs:

- ❑ They make for easy replacement in the field.
- ❑ They facilitate diagnostics.
- ❑ They are especially useful in this age of mass customization, where modularization permits postponing customization until the latest moment possible—say, at the customer's or dealer's site, or at least until final assembly and test.
- ❑ They permit parallel design work in teams, instead of series design, where the start of a particular section must wait for the completion of the previous section.
- ❑ They allow a rapid exchange of a failed module in MEOST laboratory tests, permitting such testing to resume speedily while a quick analysis of the failed module can be made.

4. Design for Manufacturability (DFM)

One of the benefits of concurrent engineering (see Chapter 8) is the influence of manufacturing inputs into the design process. Two disciplines—design for manufacturability and design for assembly—emerged twenty years ago to ease manufacturing difficulties faced by line operators in production.

Two university professors—Geoffrey Boothroyd and Peter Dewhurst—have perfected these design disciplines so that any assembly can be quantified in terms of a numerical score to measure ease of manufacturability.[21] A score of 100 is best; one is worst. Progressive companies will not allow a designer to go to production without a minimum score of 80. Here are ten basic rules for achieving best results in DFM.

Design for Manufacturability Guidelines (Simplify Before You Automate)

1. Minimize the number of parts.
2. Minimize assembly surfaces.
3. Design for Z-axis assembly (using gravity as an assistant).
4. Minimize part variations (high C_{pk}).
5. Design parts for multiuse and ease of fabrication.
6. Maximize part symmetry.
7. Optimize part handling.
8. Avoid separate fasteners where possible.
9. Provide parts with integral self-locking features.
10. Drive toward modular designs.

There is one more guideline that should be added: poka-yoke.

5. Design for Mistake Proofing: Poka-Yoke[22]

Many quality problems are blamed on the poor operator, even though quality giants, such as Deming and Juran, stress that 85 percent of such problems are the fault of management. It is human to make mistakes, even among the highest-paid athletes and CEOs. Punishing and reprimanding the assembler or adding visual inspection upon visual inspection is brute-force, illogical quality.

Thirty years ago the Japanese invented a technique—poka-yoke or mistake proofing—whereby a sensor or signal (electrical, mechanical, or plain visual) is sent to the operator when a mistake is about to be committed, so it can be prevented just-in-time. The simplest example of poka-yoke is a three-prong male plug at the end of a line cord. It will not go into a female line socket if plugged in wrong. Poka-yoke is now so widespread in Japan that it has eliminated the need for control charts, which, unfortunately, the United States has re-embraced.

Poka-yoke starts in design. In fact, 75 percent of so-called operator-controllable defects can be corrected in design, where a careless error is detected at the moment of assembly and is avoided. These workmanship problems—probably one of the major causes of infant mortality failures in reliability—are caused by the absence of poka-yoke. Attention to DFM and poka-yoke at the design stage not only enhances early life reliability, it can also speed up MEOST in the lab by eliminating needless stoppages in testing.

6. Design for Minimum Variability (High C_{pk})

One of the most important objectives of quality is to reduce variation in key product parameters. Variation is evil. Why? Generally, up until the mid-1980s, industry was comfortable having a process width (maximum process limit minus the minimum process limit) as wide as the specification width (maximum specification limit minus the minimum specification limit). This ratio of specification width divided by process width is called C_p. And industry was comfortable with a C_p of 1.0.

But a C_p of 1.0 leads to a minimum defect level of 2,700 PPM, or 0.27 percent, which is unacceptable in the twenty-first century where defect levels below 100 PPM are required and becoming commonplace. (The 2,700-PPM level applies only to a normal distribution. For the Camp Meidel approximation [skewed distribution], the defect level can be as high as 50,000 PPM—an astronomical figure by today's quality standards.)

Using Design of Experiments (DOE) techniques,[3] C_{pk} is the correction of C_p (by means of a formula) when the average of a parameter distribution is not at its design center or target value. If the average at the design center is at the same point, $C_{pk} = C_p$. If not, C_{pk} is less than C_p. The DOE world-class standard for minimizing variability is a C_{pk} of 2.0. This means that the process width of a key parameter should be no more than half the allowed specification width. With a C_{pk} of 2.0, the defect level drops to two parts per billion units (PPB). Not only is this perfect quality, but it greatly reduces scrap, repair, rework, warranty cost, inspection, test, and cycle time. No company can afford C_{pk}s less than 2.0.

The tie-in with reliability is equally important. It is well known that the higher the defect rate in manufacturing, the higher the failure rate in the field. This is not a mathematical equation or a rigid correlation, yet it exists. One of the best ways to reduce field failures is to reduce plant parametric defects and failures by forcing C_{pk}s up from 1.0 to 1.33, from 1.33 to 1.66, and finally from 1.66 to 2.0. Design of Experiments (Shainin/Bhote) is one of the easiest, simplest, fastest, and most cost-effective ways to achieve such spectacular results.

7. Design for Robustness: High S/N Ratio

In production and in the field, there are a number of uncontrollable factors, sometimes called noise factors, that can adversely affect the quality of the product. Ambient temperature, humidity, static electricity, line voltage fluctuations and transients, lack of preventive maintenance, and some degree of customer misuse can degrade design quality and reliability. Product robustness means making the product impervious to these noise factors. This cannot always be done, but robustness should be attempted in design before the product is launched into production and well before it is sent to the field.

The necessary discipline is again DOE. Using a second round of a Variables Search[3] experiment, the uncontrollable (noise) factors are deliberately introduced to determine their individual or collective impact upon the response (or Green Y). If the noise factors prove to be unimportant, they can be ignored. If they are important, some of the other product parameters must be modified so that we can live with the noise factors present. The objective is to increase the signal-to-noise (S/N) ratio by optimizing the response and minimizing the effect of the uncontrollable factors.

> Here's an example of the merits of design for robustness. An appliance manufacturer was experiencing an unacceptably high failure rate of its clothes dryer in consumer homes. The two principal reasons were 1) poor venting, where outlets clogged up, and 2) line voltage fluctuations. Both were beyond the manufacturer's control. The consumer could not be expected to worry about the venting, nor did he have control over line voltage fluctuations caused by the power utilities. A Variables Search experiment was run in production. It indicated that airflow was a parameter that could be increased with minimal cost increases, making the dryer impervious to venting constrictions and line voltage fluctuations. The company now has a "robust" consumer product with a field failure reduction of over 10:1.

Not only does design for robustness help improve field reliability by making the design "robust" to withstand the uncontrollable noise factors, it is also a way to factor these noise factors into the MEOST cycle of test.

Generally, MEOST stresses do include temperature, humidity, or voltage fluctuations and transients. But other uncontrollable noise factors, such as deliberate maintenance deterioration, customer misuse through ignorance and carelessness, and sloppy installation should form part of a MEOST regimen, if not in Round 1 of MEOST, then at least during a follow-on Round 2. Remember, Murphy's Law is omnipresent—in design, in production, and in the field.

8. Design for Maintainability and Availability

Maintainability is a companion piece to reliability. Theoretically, in a perfectly reliable world, maintenance is zero and can be ignored. In reality, some maintenance is required, even on simple products.

Two metrics are commonly used to assess maintainability:

1. Mean time to diagnose (MTTD)
2. Mean time to repair (MTTR)

Both metrics can be quantified with targeted goals. MTTD can be reduced considerably through diagnostics, especially *built-in diagnostics,* which should be factored into the design. Given the shoddy nature of the service industry and its dependence on untrained, uncaring service personnel, manufacturers must do more to design fault-finding indicators, because that will make the task easier for dealers, services, and customers.

MTTR can be reduced by 1) applying design for manufacturability (DFM) guidelines that also simplify the disassembly process; 2) using built-in diagnostics that quickly pinpoint the problem area and reduce troubleshooting; and 3) adopting modular designs that facilitate quick exchanges of modules.

MTTD and MTTR together constitute *downtime* for a product. For the average consumer, downtime is not a life-and-death issue. It may be a nuisance and an annoyance, but—for the most part—it does not affect his pocketbook. For certain industries, however, such as transportation and power generation and many other services, downtime can mean the loss of much revenue. Consider the following:

- A truck that's disabled for poor reliability can cost the trucker thousands of dollars in hauling fees, not to mention further loss due to spoilage of perishables.
- An interruption of power on a grid to a city of even one minute, due to a failure in the electrical system, such as a relay, can cost the power company over $1 million—not counting the costs to its millions of customers.
- The blackout disaster in the eastern United States and Canada in 2003 was caused by the domino effect of a single power station failure in Ohio. The total cost was in the billions of dollars.

In such cases, downtime becomes far more important than mere warranty (unless there are specific financial penalties associated with the field failures).

9. Design for Resale (Residual) Value

Most companies are not concerned with the resale value of their products. But for all capital-intensive products (e.g., planes, buses, cars, earth-moving equipment, etc.), ignoring resale value can cost the company loss of revenue and loss of profit. U.S. automakers pay warranty costs of $462 per car per year, but lose another $5,000 per car in the same original price bracket in differential between their resale value and those of German and especially Japanese cars.

Resale value is a function of cosmetic factors, public perceptions, and dealer inputs. Above all, it is a function of reliability and durability, determined in advance—at the design stage—with a MEOST purification.

10. Product Liability Analysis: Preventing Ambulance Chasing

No aspect of reliability has raised so much controversy as product liability and the lawsuits it has spawned. Consumers are fed up with shoddy, unreliable, unsafe products from unscrupulous and uncaring manufacturers. The manufacturers claim that court judgments on lawsuits filed by plaintiffs are capricious and unfair and so punitive that companies are driven out of business. Punitive damages have skyrocketed to stratospheric levels involving millions of dollars.

The brief on behalf of the plaintiffs:

- More than 20 million nonwork-related injuries each year are caused by U.S.-made consumer products.
- Of these injuries, 110,000 result in permanent disability.
- Another 30,000 result in fatalities.

For their part, the manufacturers claim that the courts have simply abandoned common sense (see the case studies for three horror stories of lawsuits and judgments out of control). Manufacturers point out that:

- Court rulings have changed from the doctrine of "caveat emptor" (let the buyer beware) to "caveat ven-

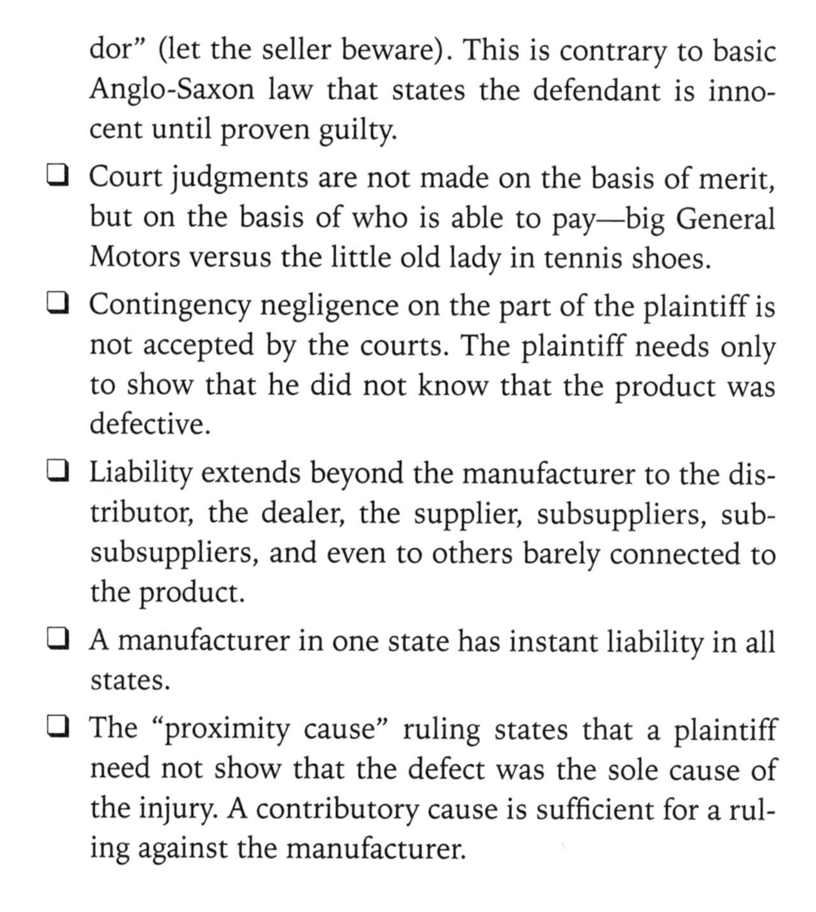

dor" (let the seller beware). This is contrary to basic Anglo-Saxon law that states the defendant is innocent until proven guilty.

- ❑ Court judgments are not made on the basis of merit, but on the basis of who is able to pay—big General Motors versus the little old lady in tennis shoes.
- ❑ Contingency negligence on the part of the plaintiff is not accepted by the courts. The plaintiff needs only to show that he did not know that the product was defective.
- ❑ Liability extends beyond the manufacturer to the distributor, the dealer, the supplier, subsuppliers, subsubsuppliers, and even to others barely connected to the product.
- ❑ A manufacturer in one state has instant liability in all states.
- ❑ The "proximity cause" ruling states that a plaintiff need not show that the defect was the sole cause of the injury. A contributory cause is sufficient for a ruling against the manufacturer.

CASE STUDY 1: COMPENSATION TO A BYSTANDER

In the case of a collision between two cars, the claim of the injured driver was amicably settled between the insurance companies. But a bystander had witnessed the collision and sued the insurance company of the driver responsible for the accident, claiming that she had been traumatized by the sight of blood on the injured driver and suffered loss of sleep, loss of sex, and loss of appetite. She was awarded $77,000 as compensation.

CASE STUDY 2: STALLED CAR ON GOLDEN GATE BRIDGE

A Ford automobile stalled on the Golden Gate Bridge. The driver was trying to start the car again when, through his rear-view mirror, he saw another car approaching. The two women in this approaching car were chatting away, unmindful of the stalled car ahead of them. The driver of the Ford jumped out of his stalled car just before the second car banged into him. On the basis of negligence, the women sued not only Ford Motor Co. and the insurance company of the other driver, but also Ford's suppliers. To avoid a lengthy litigation, which would have cost Ford and its suppliers over a million dollars, a nuisance settlement was reached for $120,000.

CASE STUDY 3: FAIRNESS OR EXTORTION?

A dairy equipment company, which also made chemical products to clean the teats of cows, was sued by a farmer who claimed that six of his cows were killed by the company's product. He wanted a settlement of $500,000. The company refused and went to elaborate lengths to prove that its product had met all safety requirements.

The case went to trial. A sympathetic jury came in with the unbelievable figure of $10 million against the company. The president of the equipment company, a man of principle, declared that he would file for bankruptcy rather than pay this new form of extortion. Eventually, a higher court dismissed the whole lawsuit.

The fear of liability lawsuits brought by ambulance-chasing lawyers need not be a permanent setback. Customers and supplier companies can and must strengthen their designs to protect themselves against these predators. Courts and juries come down much harder on a design deficiency when it results in personal injuries and damages rather than manufacturing problems alone. A design deficiency is likely in 100 percent of a product; a manufacturing deficiency in only a small percentage of the product.

Table 9-1 provides guidelines that will greatly reduce, if not completely inoculate, a company against lawsuits. Above all, document all your actions in product liability prevention. Courts are impressed with such documentation and proof of intent.

Product Liability Analysis (PLA) Example

PLA is similar to a fault tree analysis (FTA), except that a PLA starts with a product failure that endagers safety. Figure 9-3 is an example of an ignition amplifier whose failure can cause a car to go dead or become intermittent. The numbers in brackets are the probability of failure.

Each failure is then assigned causes, which in turn generate subcauses. The critical path is the path that ends up with the highest probability of failure and should be worked on the earliest. In Figure 9-3, the critical path is:

Car intermittent (70%) → Wire bond on power transistor (50%), with a risk of probability of 70% × 50% = 35%

(text continues on page 144)

Table 9-1. Product liability guidelines for engineers.

Definitions

Hazard	A condition with potential injury.
Risk	The percent probability of injury from a hazard.
Danger	The combination of hazard and risk. To reduce danger, either hazard or risk must be reduced.

- ❑ Assume that design will go to trial.
- ❑ Assume that every design decision will be scrutinized in a court.

Duties Imposed on Engineers for Product Liability

1. The duty to design for all foreseeable hazards—including guarding against foreseeable misuses of the product and minimizing risks to users (and nonusers) if accidents are unavoidable.
2. The duty to investigate and test to discover risks—including investigation and correction of field problems.
3. The duty to minimize risk by alternate design, including safety device warning. Courts do not insist on completely fail-safe designs; "reasonably safe" is adequate. In this regard:
   - ❑ The designer may reject a safer alternate design if it is not feasible, is unduly expensive, or involves other dangers.
   - ❑ Because warnings are inexpensive and practical, courts have even greater expectations of the efficiency of warnings than they do of alternate designs and safety devices.
   - ❑ Warnings must be prominent enough to attract user attention and clear enough for users to understand the nature of risk.

4. The duty to meet government/industry safety standards:
   - ❑ If a government standard is not met, the manufacturer is liable per se, without an opportunity to defend the design.
   - ❑ If an industry standard is not met, or is less safe than competitive products, the manufacturer is not liable per se, but will be found liable by a jury.
5. The duty to report field defects:
   - ❑ Under the Consumer Product Safety Act, a manufacturer of consumer products must report to the U.S. Consumer Product Safety Commission (CPSC) if a defect can create substantial risk of injury to the public. Penalties for nonreporting are severe.

Successive Lines of Defense

1. Avoid the accident.
2. If a hazard cannot be eliminated for functional reasons, protect against it.
3. Make the accident "safe" by designing for minimum damage.
4. Predict or warn of impending accident.
5. Warn of possible (versus impending) accidents.
6. Protect the user in case of accident.

Trade-Offs: Safety vs. Cost and Safety vs. Function or Utility

- ❑ The objective should not be safety over other parameters but an optimization and elevation of all factors.
- ❑ Safety must be a specification, along with cost, performance, and reliability.

Product Liability Prevention Analysis

1. Assign probabilities and severities of failure to each branch, down to the root cause, where:

$$\text{Probability} \times \text{Severity} = \text{Risk Priority}$$

2. Determine critical path (i.e., highest products of risk priorities).
3. Conduct a failure mode and effects analysis (FMEA).

MEOST to Assist Product Liability Detection

The beauty of MEOST is its speed in generating failures. It can, therefore, play a significant role in "seeding" potential failures, such as the potential for customer misuse, in order to see their effect on the product—yet another new frontier for MEOST.

11. Destructive Physical Analysis (DPA)

A reliability discipline that is seldom practiced in industry is to take apart components or subassemblies received from suppliers and examine them for potential violations of layout or construction, and for substandard manufacturing practices. Electrical tests alone may not detect such anomalies. DPAs can also be performed on finished product. The sample size is generally one or two units, performed once a week or once a month. Many field problems can be detected through DPA long before they are finally apprehended in the hands of customers months or years later. DPA can also prevent failure-prone components getting into MEOST tests.

Figure 9-3. Product liability analysis—ignition amplifier.

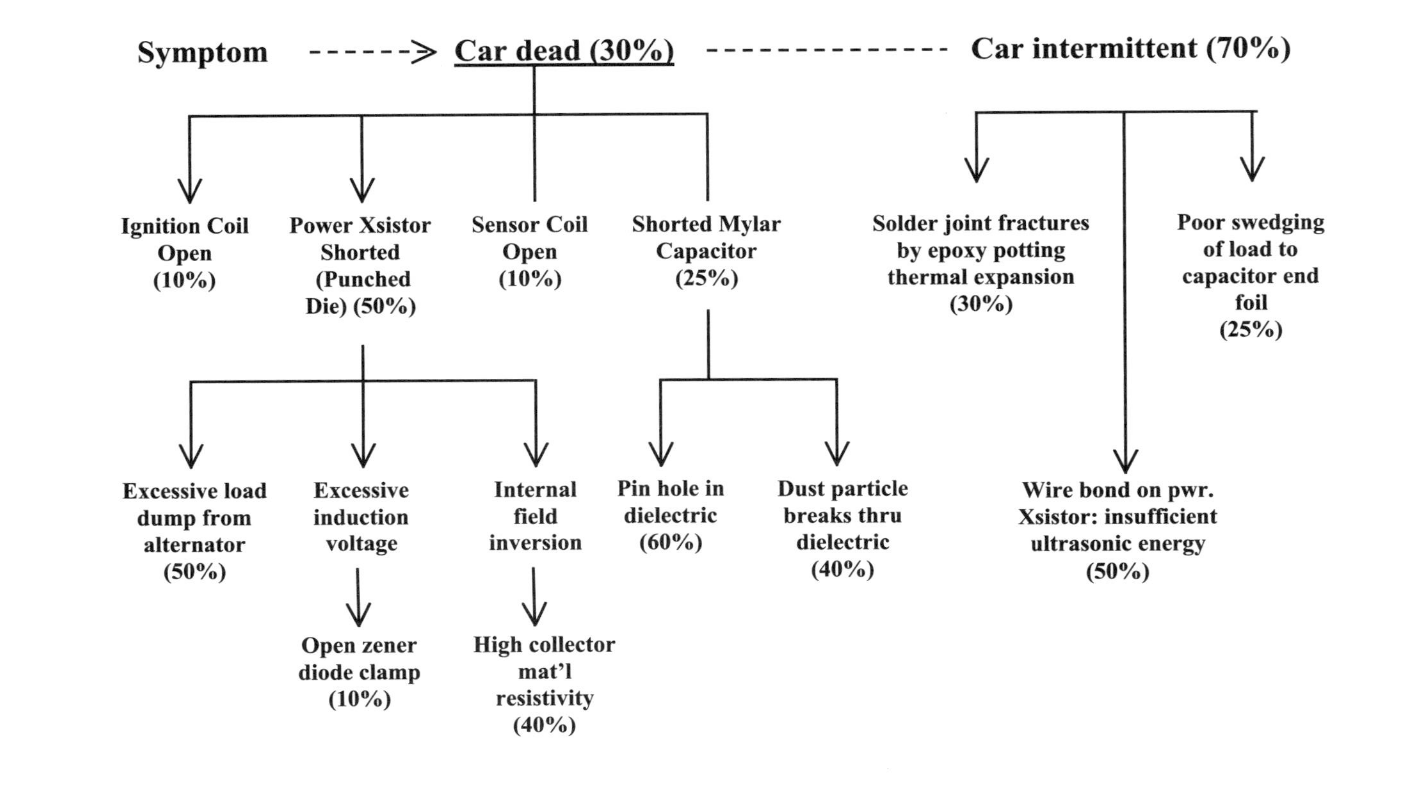

12. Field Escape Control

In a production plant, DPA can be considered the first line of defense against poor field reliability. The second line of defense is field escape control, which scrutinizes line failures with reliability implications. It stands to reason that if reliability-oriented line failures are not acted upon, they are bound to repeat themselves in the field. The factory floor thus becomes the largest and earliest of a company's service stations.

Eighteen years ago, we introduced field escape control at Motorola as an early-warning system for later field failures. The rules were:

- ❑ Separate the failures found in all testing into two groups:

  Category 1: Quality-related failures that are unlikely to be passed on to the field

  Category 2: Reliability-oriented failures that are likely to be repeated in the field

- ❑ If there is only one reliability-oriented failure mode for each component per day, treat it along with the quality-related failures.
- ❑ If there are two or more reliability-oriented failures for each failure mode per day, treat them as crises and dispatch them to failure analysis, then correct the failure as soon as possible.
- ❑ Continue monitoring future line-test failures in category 2 to ensure the reduction or elimination of these reliability-oriented failures.

13. Mini-MEOST Sampling in Production

Burn-in is still fashionable in many electronic companies. This practice is often demanded by unthinking customers as a way to ensure reliability and weed out infant mortality or early-life failures in the field. Typically, the units are subjected to a high-temperature soak (usually not more than 85° C), with or without power applied to them for periods ranging from twenty-four hours up to ninety-six hours. Burn-in is a 100 percent test.

There are, however, many disadvantages and very few advantages to burn-in:

- ❑ It detects the most superficial defects, and in any case, not more than 5 percent of potential field failures.
- ❑ Most regrettably, it consumes precious cycle time and throughput time in production. Today, with the emphasis on inventory turns, companies are beginning to reduce manufacturing cycle time from one week to one day to one hour. Burn-in needlessly adds one to four days of cycle time, negating all the gains of cycle time reduction.

If a company is still wedded to burn-in, a better method would be to do burn-in on a sample test basis, rather than as a 100 percent test.

Mini-MEOST Practices

Instead, it is highly recommended that 100 percent burn-in—or, for that matter, HASS test—be thrown out and a truncated MEOST or mini-MEOST be used instead. This would be a repeat of MEOST in design (see Chapter 10), but modified as follows:

- ❑ Only the most important environments/stress would be used, instead of the larger number of stresses in design. Typically, temperature cycling and vibration, or temperature cycling and humidity, might suffice.
- ❑ The stress levels would be at the operating level (described in Chapter 10) instead of the higher Maximum Practical Over Stress Level (MPOSL).

Mini-MEOST would be done only on a small sample of 5 to 10 units, once a week or once a month, depending on production volumes. The purpose is to ensure that the integrity of the design, established in engineering, is not degraded by suppliers, workmanship, and production processes. Mini-MEOST can, therefore, be considered as the third line of defense in reliability production—the first being DPA and the second, field escape analysis.

14. Dealer Feedback and Zero Time Failures

A fourth line of defense is getting fast and accurate feedback of failures at distributor/dealer centers. When large products such as earth-moving equipment, heavy-duty trucks, and even automobiles are involved, the dealer is the first location where all assemblies come together before the "prepping" is done or the installation is finalized. With the popularity of mass customization, the final customization (or options) is postponed until the total configuration comes together at the dealer site. Feedback and corrective action can be installed before the customer sees the product for the first time.

Perhaps a fifth line of defense is to get feedback from the customer on virtually zero time field exposure. These zero time failures—or dead-on-arrival (DOA) failures—are the early harbingers of future trouble in the field. If a hard-hitting corrective action is instituted, it may fend off future trouble that would otherwise continue unabated after months and years of service. The failure profile can then be compared against the profile of failures found in design MEOST.

15. Field Failure Reliability

A major weakness that is common to most companies is the inaccuracy in gathering field failure data. The causes of poor field data retrieval are:

- Customers who cannot describe the failure in specific terms
- Service writers who are too busy or too untrained in interpreting the customers' complaints
- Service stations and servicers who are poor at diagnosis, exchange parts willy-nilly to locate the problem, and charge the manufacturer for all of them
- Dishonest service stations that turn in fictitious claims to make a fast buck
- "Weekend" customers who buy a product, use it, and return the product as defective a few days later

A company is not helpless in overcoming these inaccuracies. It can, for instance:

- Set up its own service stations—rather than depend on external services—to get accurate field reliability data

- ❑ Establish "listening posts"—that is, a few large and reliable service stations—to get early and dependable field failure feedback
- ❑ Call the customer directly to verify the exact nature of his complaint and to assess his satisfaction with the repairs performed

Scheduled Servicing

There is a tendency among companies in their own plants and among customers of their products to ignore scheduled servicing. Under this practice, a product is checked at periodic intervals to remove critical parts or subassemblies before a specified age or time, in order to substantially reduce the likelihood of failure. Instead, their rationale is, "If it ain't broke, don't fix it."

Scheduled servicing is now standard practice in the automobile industry. The question, however, is: How do you determine the length of time (or miles) between successive servicing? A research study conducted on previous field maintenance practices found that 89 percent of the items that had been designated for removal had no decrease in reliability with age.

MEOST studies can be conducted to determine the optimum age or time required for removal and replacement of a critical part, so this determination can be made at the design stage, instead of waiting for months of field experience.

Reliability of the Maintenance Function

But how reliable is maintenance itself? An Air Force study found that 2 percent to 48 percent of the total failures of electronics equipment were caused by poor maintenance. On the other hand, consider the thoroughness of General Motors

Chevrolet division's maintenance, which identifies twenty-two areas essential for its customers' total satisfaction. For each of the areas, specific evaluations are used to determine if a dealer is worthy of maintenance certification. This includes a review of the dealer's procedures and facilities as well as customer feedback on dealer performance.

Spare Parts

Companies may lose money or break even on their initial sales to customers, but they make up for the loss with bumper profits in spare parts. One large truck company lost more than $1 billion in sales, but its parts division turned in a handsome 32 percent return on investment. That is the wrong way to go. Traffic in part sales is a reflection of poor reliability, for which the company will pay with a high rate of customer defections. In the case of the truck maker, the company lost 8 percent of its customer base, which includes some of its most profitable customers.

At the same time, if maintenance is necessary, so are spare parts. There are two conflicting requirements. On the one hand, spare parts must be available to dealers/servicers within a matter of days. (Caterpillar, Inc., the benchmark in this area, guarantees that it will deliver a needed part anywhere in the world in no more than forty-eight hours.) On the other hand, keeping a large inventory of spare parts is costly, especially given a governmental rule that spare parts should be available for at least seven years after the product is obsolete in production.

Again, the dilemma can be solved with MEOST. In the design laboratory where MEOST is run, critical parts that are likely to fail—and when—are identified. First, their reliability should be improved. Second, provisions should be made to have an adequate supply of such parts for service. The result:

both maximum customer satisfaction and minimum overall inventory of spare parts.

The Human Error Factor

Several studies have been made to identify the extent and effect of human errors on product reliability and maintainability. Up to 50 percent of field failures in major systems can be assigned to human errors, of which there are several types:

- Failure to allow for training in how to operate the new product.
- Failure to use available information—Somehow it is "macho" to ignore the owner's manual and instructions, except as a last resort.
- Use of the product in applications never intended—An extreme example was a customer using a lawnmower as a hedge trimmer, and then suing the company for its failure to warn him when his leg was cut off!
- Failure to follow scheduled maintenance—Consumers are notoriously lax in following a prescribed schedule for lubrication, cleaning, and replacement of parts.

In many cases, prevention of human errors in use should start with product design, just as the prevention of operator-controllable defects in production—poka-yoke—starts with design. This includes ergonomics, built-in diagnostics, and fail-safe features. The use of service contracts for proper maintenance can reduce warranty and add to business profitability. As examples:

- A clothes dryer manufacturer found the clogging of vents to be a perpetual problem in users' homes. It offered a periodic vent inspection and correction to the homeowner for a small fee.

- A milking equipment company was getting complaints from its dairy farmer customers about its milk lines being responsible for their receiving lower milk prices, based on a higher bacteria count. The company offered a free audit of the entire installation and ended up taking 60 percent of the business away from its competitors.

Longer Warranties

With attention paid to good data retrieval, scheduled servicing, maintenance reliability, spare parts, and human factors, the stage is now set to accurately track field failures, including the figures from three months in service, six months in service, and twelve months in service. There are a few facts to remember:

- In calculating failure rates, there should be a time offset between the time of shipment to the field and the maturing field data. In other words, the shipment figures would be for the previous three months (sometimes six months) to allow the production units to actually start being used by customers.
- Generally, field failure rates decrease with time, even beyond the infant mortality period. There is no such thing as a constant failure rate.
- The one-year failure rate should be approximately twice the three-month failure rate, and a two-year failure rate approximately twice the six-month failure rate.
- Beyond the warranty period, all field data becomes very murky and subjective. Only a MEOST Weibull extrapolation

(as explained in Chapter 10) can project failure rates up to five, ten, and twenty years.

- It would be desirable to form partnerships with large and key customers who could track field failure data beyond the warranty period in exchange for some type of financial or nonfinancial incentives.

CHAPTER 10

The Climb to the Mount Everest of Reliability—MEOST

"Why did I choose to climb to Mount Everest? Because it has been a feat unsurpassed in 4,000 years of mountain climbing. . . ."

—SIR EDMUND HILLARY

The Salient Principles of MEOST

Although some of the principles of MEOST have been stated in earlier chapters, it is important to highlight them together.

1. In MEOST testing, it is not the objective to pass a product, but to fail it. It is only through failures that the weak links of a design can be "smoked out." Failures—paradoxically—mean success.
2. A single stress/environment is not enough to generate failures.

3. More than one stress applied sequentially to the product is also not enough, because interaction effects are not detected.
4. Several stresses/environments must be combined, duplicating, as closely as possible, field conditions to generate failure synergy through interaction effects.
5. The combined stresses must go well beyond design stress levels to a Maximum Practical Over Stress Limit.
6. The rate of overstress must be accelerated to produce failures in the shortest time possible.
7. Weak parts have induced stresses that are two and three times more severe than good parts, and they can be made to fail with higher rates of stress in a very short time.

Miner's Equation

The underlying theory behind principles 5 and 6 is contained in a famous law called the Miner's equation:

$$D \alpha \times n \times s^{10}$$

where:

D = Damage Accumulation or Failure

α = Proportional to

n = Number of Cycles

s = Stress

Of all the reliability models, Miner's equation comes closest to real-life situations. It means that the degree of stress has ten times (or more) the ability to smoke out failures than the number of cycles (usually thermal cycles). The equation enables us to dramatically shorten the MEOST test time by accelerating stress levels. Miner's equation works well for stresses such as voltage, current power, speed, humidity, and common concentrations.

Figure 10-1 shows how the traditional bathtub curve shortens with each increasing stress and simultaneously increases the failure rate.

Figure 10-2 shows the effect on damage (failures) caused both by the number of cycles of stress and the magnitude of the stress. Increasing stress accelerates failure much more than increasing the number of cycles.

Figure 10-3 is a dramatic portrayal of how increasing the rate of temperature shortens the time (or cycles) to failure. A rate of 5° C per minute in temperature would require 400 cycles to achieve the same failure as 25° C per minute would achieve in four cycles—a 100:1 time compression. It is no

Figure 10-1. Bathtub curve showing stress vs. expected life relationship.

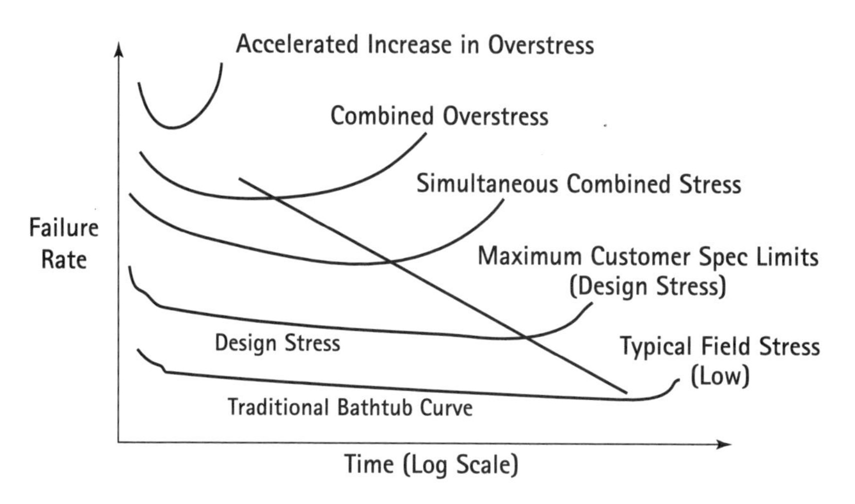

Figure 10-2. Stress levels vs. number of cycles.

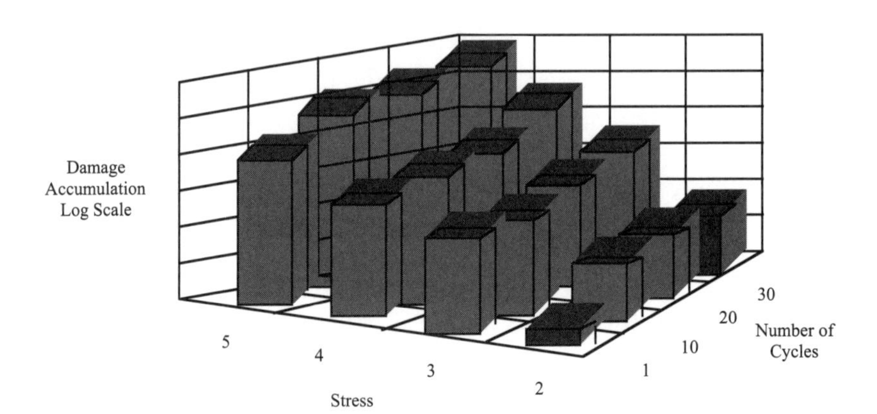

wonder that older chambers with rates of temperature change of 2° C to 3° C per minute could hardly generate failures in the old days of MEOST. Similarly, for mechanical stress, a doubling of stress could reduce the number of fatigue cycles required for failure by 1,000:1.

Figure 10-3. Thermal cycling rate vs. number of thermal cycles.

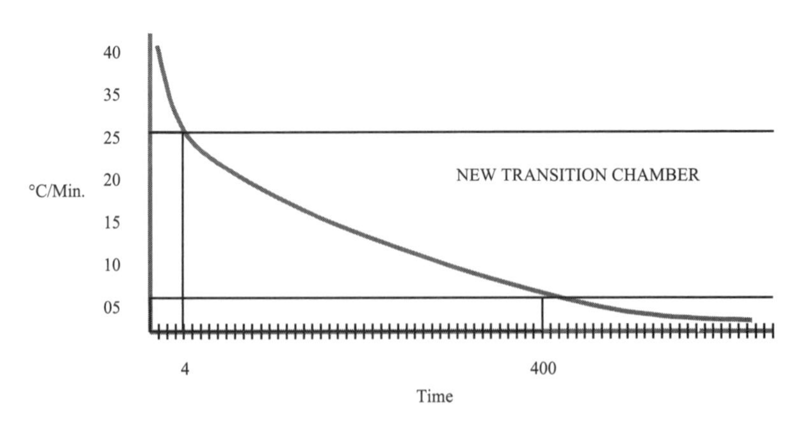

Source: Steve Smithson, *Effectiveness and Economics—Yardstick for ESS Decision.*

MEOST History, Applications, Successes

History

In the 1950s and 1960s, most U.S. space shots would flop into the Atlantic Ocean, while Soviet spacecraft soared into the heavens. In the Apollo series of space shots, every mission had failed except one—the Lunar Module that took two astronauts up to the moon and back (see Figure 10-4). Had the Lunar Module failed, we would have had the moon overpopulated by two persons! But it didn't fail. The Apollo Lunar Module had been subjected to the first MEOST methodology.

Applications

Since the 1960s, MEOST has been successfully applied to aircraft engines, helicopters, automobiles, motorcycles, railroad cars, buses, water treatment systems, air treatment systems, electronics, aerospace equipment, and several other products.

Motorola Successes

At Motorola, we first applied MEOST to an engine electronics module, which had a failure rate in Ford cars of one percent to 2 percent per year. Ford was unhappy with this reliability (even though the module had passed Ford's elaborate but ineffective single environment sequential testing), and so were we.

With a simple preliminary combing of environments, even without any overstress, we found failure modes, in our first attempt at MEOST, that matched the principal failure modes in the field. We continued refining our MEOST test plan and corrected the failure modes. The final result—a 0.01 percent failure rate/year, which is more than a 100:1 improvement.

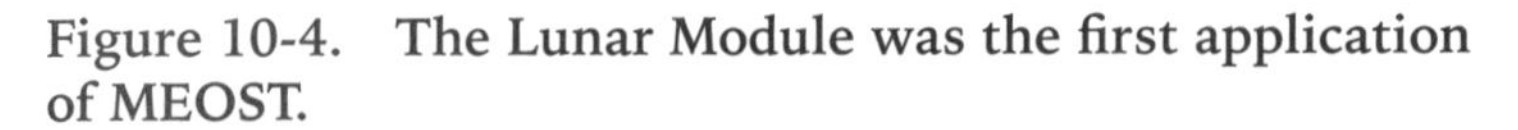
Figure 10-4. The Lunar Module was the first application of MEOST.

Source: NASA.

Probably our most notable success at Motorola was on an electronics module for the Cummins Engine Company. The reliability specification was that our modules had to last for 200,000 miles on their engines. Our MEOST process unearthed several design failures before the unit went into production. The final result: The product survived seven years in the field without a single failure. And some of those engines had chalked up over 500,000 miles.

Since the early days, Motorola's electronics products, subjected to MEOST, have registered failure rates per year of less than 10 parts per million (PPM). As other examples:

- Motorola provides all radio links on all of NASA's space probes to the outer planets of our solar system. With the help of MEOST, these space probes have traveled millions and millions of miles for more than eleven years without a single failure.
- Motorola's venture into global satellite communications was Iridium. This sixty-six satellite system was successfully validated with MEOST. (It turned out to be a technological marvel, but a financial flop for Motorola.)

Automotive Successes

Failure mode verification tests (FMVT) is a variant of MEOST. FMVT takes longer to test because it continues stress testing until the failures are bunched at very high stress levels and have to be analyzed at needless expense. There are, however, successful applications of this technique in the automotive industry, as noted in the case studies.

CASE STUDY 1: CAR DOOR[23]

Table 10-1 shows the difference in validation test time, labor time, test cost, prototypes required, and failure modes uncovered between traditional engineering test and FMVT/MEOST tests when the product is a car door.

Table 10-1. Case study 1: car door.

Parameter	*Traditional Eng. Test*	*FMVT/MEOST*
Test Time	172 Hours	16 Hours
Labor Time	117 Man-Days	38 Man-Days
Test Cost	$124,000	$44,000
No. of Prototypes Required		
• Vinyl	50	2
• Sheet Metal	16	2
• Savings	—	$240,000
Failure Modes Discovered	1	15

CASE STUDY 2: CAR COCKPIT MODULE

Table 10-2 shows the difference in validation time, validation cost, and potential impact on warranty between traditional engineering tests and FMVT/MEOST tests when the product is a car cockpit module.

Table 10-2. Case study 2: car cockpit module.

Actuation Fixtures: Accelerator Pedal; Break Pedal; Clutch Pedal; Power Steering; Door Jamb Switch; Glove Box Door; Hood Release; HVAC Outlets.

Parameter	*Traditional Eng. Test*	*FMVT/MEOST*
Total Test Time	653 Days	225 Days
Total Cost	$1.95 Million	$1.32 Million
Failure Modes Discovered	21	62

Note: FMVT goes much further in overstress than MEOST. It takes longer and is more expensive and less effective than MEOST. Furthermore, many FMVT failures are not truly relevant.

Types of Stresses: Combined vs. Isolated Stresses

It is not always apparent how many stresses/environments can impinge on a wide range of products. For space travel, the list is almost endless and very hostile. For transport (on- and off-road) vehicles, the list is almost as formidable. For household appliances, which most people think operate under benign conditions, the number of stresses can be surprising. Table 10-3 is only a partial list of such stresses for each category of product.

A cardinal principle in MEOST is to combine these stresses to detect interaction failures. But there is no test chamber large enough or complete enough to accommodate all or even most of these stresses.

Before designing a MEOST process, a decision must be made on which of these stresses are likely to operate simultaneously on the product and which are likely to operate in isolation. As an example, if the end-use of a product is stationary, vibration and/or shock may only be important during handling and transportation. Under such circumstances, vibration can

Table 10-3. Typical stresses/environments for three categories of products.

Outer Space	*Transport Vehicles*	*Washing Machines*
Gravity	Temperature Cycles (Summer to Winter)	Temperature Cycles
Meteors	Temperature Cycles (Day to Night)	Humidity
Cosmic Ray Bombardment	Humidity	Chemical (Bleaches, Detergents)
Sunspot Flare-Ups	Shock	Over & Under Loading
Radiation	Vibration	Fabric Fibers
Atmospheric Absence	Electromagnetic Compatibility	Lint
+ Most Items in Column 2	Voltage Transients	Sand
	Load Dump	Water Hardness
	Salty Spray	Water Pressure
	Dust	Line Voltage Variations/Spikes
	Oils	Floor Levels
	Gasoline	Foreign Objects
	Chemicals	Dirt
	Snow, Hail, Rain	Flood Residue

be tested separately. (Vibration adds significantly to the complexity and cost of test chambers.)

In addition, of the stresses likely to operate simultaneously, only four or five of the most important can be factored into a MEOST arrangement from a practical, cost-effective point of view.

Design Stress vs. Operational Stress vs. Maximum Practical Over Stress vs. Destruct Levels

There are three stress levels folded into design stress:

1. Engineering specifications for each stress level
2. Customer requirements for each stress level
3. The maximum stress level likely to be reached in the field under the most severe geographical and climatic considerations

Of these, the hardest to define are the maximum stress levels seen in the field. Several companies equip trucks with instrumentation to measure the stress levels of, say, temperature in the Arizona desert in summer and the Canadian Arctic in the winter. Or they may measure other extremes, such as humidity in Louisiana's bayou country and salt spray near the Florida Keys.

For purposes of MEOST:

- ❑ *Design stress* is defined as the highest of the aforementioned three stresses.
- ❑ *Operational stress* is defined as that level of stress, in the overstress region beyond design stress, when failure first occurs, but when it is reduced, the product recovers.

- *Destruct stress* is defined as that level of stress that causes failure, but when it is reduced, does not allow the product to recover. A second definition is a stress level where failures occur very rapidly. A third and more severe level of destruct test is defined as the fundamental limit of technology (FLT), where the product completely falls apart. As an example, the FLT for solder is its melting point—around 255° C; the FLT for plastic when it softens is around 100° C.

Figure 10-5. Typical stress levels.

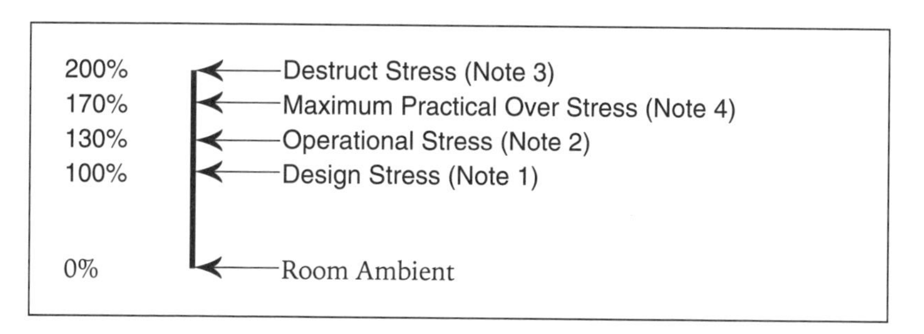

Note 1: Design Stress is the highest of three stresses/environments: 1) Eng. Specs, 2) Customer Requirements, 3) Max. Field Env.
Note 2: Operational Stress is that stress which, when reduced, causes the failure to be healed (a HASS level).
Note 3: Destruct Stress is 1) that stress level which, when reduced, causes continuity of failure (a HALT level); 2) the stress level at fundamental level of technology.
Note 4: Maximum Practical Over Stress is a stress level midway between operational and destruct levels (a MEOST level).

Figure 10-5 shows the relative levels of the three stress levels—design stress (the lowest stress), which is the starting point of overstress; operational stress; and destruct stress. As a rule of thumb, the operational stress is one-third of the distance between design stress and destruct stress, but it is closer to design stress. We also see that:

- In HALT, the overstress is pushed right up to destruct levels, causing all kinds of failures—artificial and otherwise. It is a definite case of overkill at great expense.
- In HASS, the overstress is taken up only to operational stress, which is acceptable. But HASS, as a 100 percent test in production, is almost as bad as burn-in.
- In FMVT, the stress is pushed almost to the destruct level. The numerous failures that may never be duplicated in real life make FMVT almost as much overkill as HALT.
- In MEOST, the Maximum Practical Over Stress Limit (MPOSL) is one-third the distance from destruct test, and two-thirds the distance from design stress. It is the most sensible level of overstress.

Preamble and Preparations for MEOST

A great deal of planning and preparation work should precede a MEOST launch. The steps are outlined as follows:

1. Choosing Appropriate Product Levels. Select the appropriate product level for MEOST—systems or subsystems or assemblies or parts level.

In general, starting at the systems level is preferable because all subsystems, modules, and parts are tested simultaneously and the interactions assessed. However, there are drawbacks:

- There can be a physical limit if the test chambers cannot accommodate the large size of the final product.

- ❑ Another limitation is that not all the subsystems and parts would be subjected to adequate overstress. Many practitioners of MEOST prefer testing at the subsystem or module level for these reasons:

___ Testing at the subsystem, module, or subassembly level is physically easier due to test chamber size limitations.

- ❑ Furthermore, more units can be accommodated than at the system level.
- ❑ The limitation is that interaction effects between two or more subsystems are lost.

___ Testing at the parts or black-box level should, generally, be left to key suppliers, who should be encouraged to use MEOST for their own economic advantage (particularly if reinforced with financial reliability incentives/penalties).

2. *Prioritizing Current Field Failures.* List and prioritize field failures on similar products. A success factor in MEOST is the ability to reproduce, in hours, the same failure modes that would otherwise take months and years to occur in the field.
3. *Ruggedizing a Product for MEOST.* Before a product can be successfully tested, you need to strengthen a product (i.e., ruggedize it) in order to minimize failures that would cause avoidable failures in MEOST and needlessly lengthen MEOST testing. The precautions include:

- ❑ Derate all critical components by at least 40 percent below their supplier-recommended levels.
- ❑ Run a Variables Search experiment to minimize the "noise" effects of uncontrollable factors, such as tem-

perature, humidity, airflow, and the potential for customer misuse.

- ❑ Do a process scrub, followed by a process certification, to ensure that all peripheral quality issues have been nailed down.
- ❑ Secure all PC board assemblies on a motherboard or card cage by using positive locking devices such as brackets or ear latches.
- ❑ Avoid sockets for ICs soon after initial code stabilization.
- ❑ Secure all screws and bolts with a torque wrench or adhesive.
- ❑ Keep high-weight components away from a PC board center, to avoid maximum deflection during vibration.
- ❑ Secure all low-mass, low-voltage vertically mounted electronic capacitors. They should be clustered together and secured with RTV silicone so that they move as a single mass. (This same cluster principle has made the Sears Tower—one of the tallest buildings in the world—such an architectural marvel.)
- ❑ Use a high-grade, high-temperature insulation wire for interconnects and cable harnesses.
- ❑ To the extent possible, avoid connectors in the design. They are one of the weak links in product reliability.

4. *Selecting Appropriate Stresses in Combination.* Remember that a main objective of MEOST is to combine stresses in order to detect interaction failures. To this end:

- ❑ List all stresses likely to impact the product in the field.
- ❑ Separate those stresses that are likely to operate in combination from the ones likely to operate in isolation.

- ❑ Select four or five of the most important combination stresses for the MEOST plan.

5. *Determining Design, Operational, MPOSL, and Destruct Limits.* Here is a simple rule of thumb:

   - ❑ Divide the distance between design limits and destruct limits into three parts.
   - ❑ The operational level is one-third of the distance from the design limit.
   - ❑ The Multiple Environment Over Stress Level (MPOSL) is one-third the distance from the destruct limit.

6. *Determining the Number of Stress Levels.* Select stress levels both below and above the design stress level, as follows:

   - ❑ *Below the Design Stress Level.* Select two, three, or four levels going up from a room-ambient level to design stress. (Some companies skip these low and intermediate levels and start right away at design stress if they are confident of the product's robustness.)
   - ❑ *Above the Design Stress Level.* Select four or five levels going up from the design stress level to MPOSL. (It is preferable, in thermal cycling, to start at the cold level and work up to the hot levels.)

7. *Allowing Enough Dwell Time at Each Stress Level.* A sufficient amount of time should be allowed at each stress level for temperatures to reach the right level inside the product. In most applications, the dwell time required is no more than ten minutes. If a large mass of product is involved, longer dwell times—up to thirty minutes for each stress level—may have to be accommodated. In the final analysis, the

overall objective would be to get as many cycles of stress operating on the product in eight hours. So longer dwell times should be shortened with blowers or turbulence inside the MEOST chambers.

8. *Establishing a Combined Stress Scale.* Once the design levels, operating levels, MPOSL levels, and destruct levels are determined in step 5, it becomes necessary to develop a single stress scale that can combine all the individual stresses. This is best done as percentages of stress levels. As a general rule:

    - ❑ Design stress is called a 100 percent stress level.
    - ❑ Operational stress is called a 130 percent stress level.
    - ❑ MPOSL is called a 170 percent stress level.
    - ❑ Destruct is called a 200 percent stress level.

9. *Preparing the Stress Sequencing Roadmap.* This is one of the most important steps in MEOST planning—the exact time when each stress should be stepped up and down in sequence with other stresses.

The X-axis is the timeline. The Y-axis is divided into four or five sections, one for each stress, such as temperature, vibration, or voltage transients, among others. The sequence of the start of each cycle of stress, the dwell time, and the completion of the cycle are plotted. An attempt should be made to accommodate at least three or four such complete cycles of combined stresses in one eight-hour period, if possible.

Figure 10-6 is an example of such a sequence for one of Motorola's early MEOST trials performed on an engine control module for the Ford Motor Company. Figure 10-7 is another example of the stress sequence of a MEOST trial, this time on a product that is part of a comprehensive protection system for the power utilities.

Figure 10-6. Multiple Environment Over Stress Test plan for simultaneous multiple environments.

Simultaneous Multiple Environments:
TEMP., CYCLE, HUMIDITY, VOLTAGE, TRANSIENTS

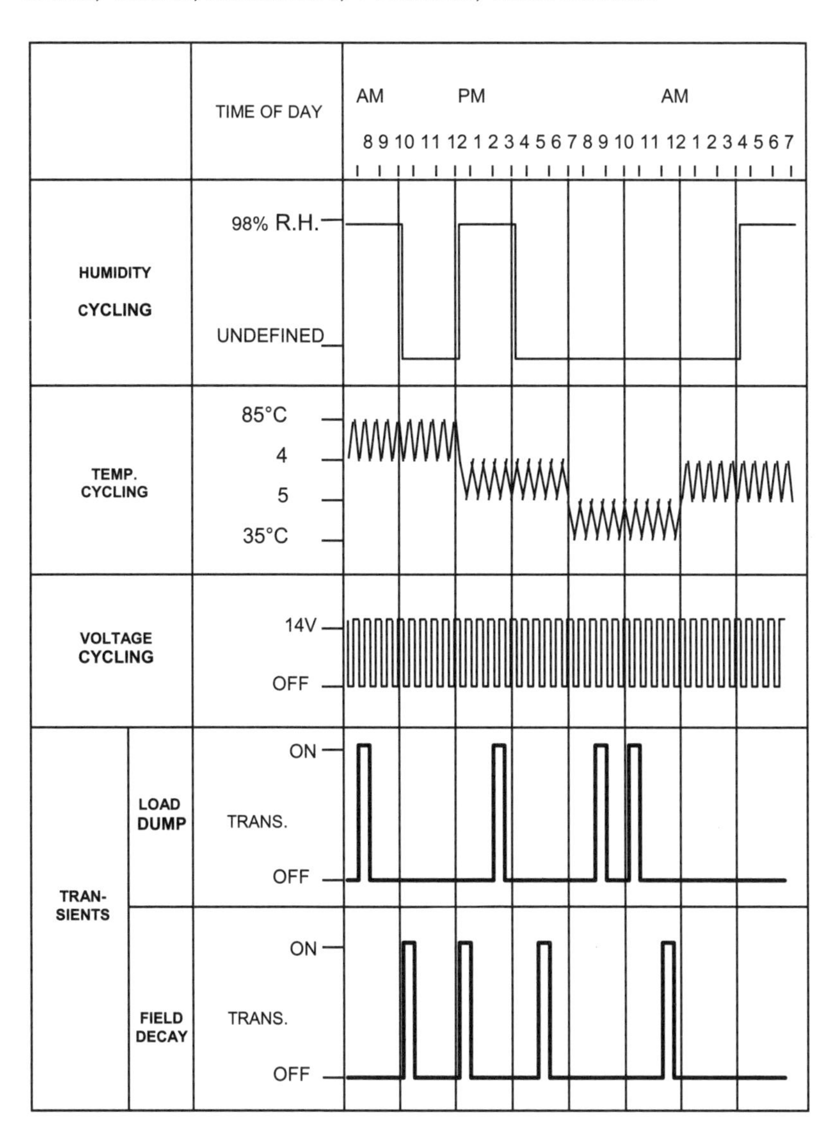

Figure 10-7. MEOST test profile.

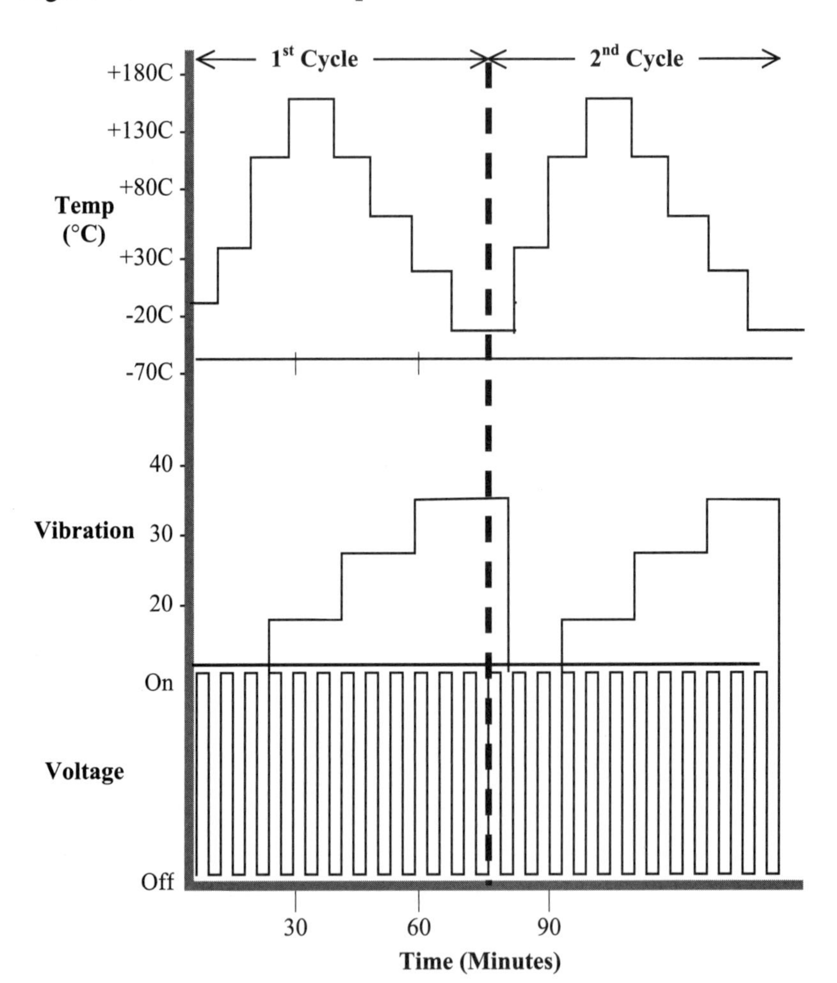

10. *Determining Outputs (Green Ys)*. A product has several output requirements that measure desired performance, such as speed, voltage, and current. These outputs or responses are called Green Ys (the lexicon used in Design of Experiments).

    A list of such Green Ys should be prepared, along with the appropriate measuring instrument. An important rule is that the accuracy of the measuring instrument must be at least five times the specification width of the parameter (or Green Y).

11. *Setting Up Adequate Support Equipment.* The following peripheral equipment should be ready before the actual MEOST launch:

    - ❑ Thermocouples to measure temperatures in a few key areas of the product
    - ❑ Accelerometers to measure vibration parameters
    - ❑ Graphical recorders to continually monitor each important Green Y
    - ❑ Data loggers for automatic data collection
    - ❑ Spare parts/modules for quick replacements after failure in the MEOST trial

12. *Choosing Sample Sizes for MEOST.* The recommended guidelines are as follows:

MEOST Stage	Sample Sizes
Prototype (stage 3)	3 for repairable units; 5 to 10 for nonrepairable units
Pilot run and subsequent stages	5 to 10 for repairable units; 15 to 25 for nonrepairable units

One of the most frequent doubts expressed by managers and engineers exposed to MEOST for the first time is: How can the small (no, tiny) sample size of 3 to 10 units in MEOST testing adequately represent the total population of the product? After all, statistics tell us that sample sizes of 30 to 50 to 100 units are a minimum required.

The answer is that, in MEOST, we are not concerned

with failures as a percentage of the total number of units tested. We want to probe the weakest components that, by the laws of physics, have stresses two to three times higher than the stronger components and, therefore, are likely to fail the quickest. Under normal field conditions, it would take a long time for even these weak components to fail. However, Miner's equation indicates that for every doubling of stress, the failure rate can jump 2^{10} or 1,024 times. As a result, the distribution of failures for several components, which may be bunched up under benign field conditions, are going to spread out under accelerated stress. And there will be a wide separation between the early failures of the weak components and the late failures (or no failures) of the more robust components.

(In MEOST stages 3 to 8, *one* failure per failure mode is allowed in the overstress region. The reason is that a single, lone failure may represent an anomaly [or maverick, a la Ford!]. As a chance occurrence, it represents an extreme low end of the failure distribution of that component and can be ignored. But *two* failures of the same failure mode is the start of a trend and should be analyzed and checked.)

The Eight Stages of MEOST

We now come to the pièce de résistance in our pursuit of reliability—the actual MEOST launch. It has eight distinctive stages.

Stage 1: Single Stress Up to the Design Limit

Purpose

This is a preliminary stage to determine the failure contribution, if any, of each single stress selected in the four or five stresses that will eventually be used in combination. Stage 1 need not be repeated for similar products. It is only necessary when there is a new product platform or new environments to be encountered. Stage 1 proceeds as follows:

- ❑ Step-stress in three or four stress levels, from room-ambient benign stress up to the design limit for that stress.
- ❑ Start with thermal cycling—say, from –20° C to 80° C, at a rate of 40° C per minute—applying a typical dwell time of ten minutes at each step and starting with the cold then hot cycle.
- ❑ For vibration—usually the second single stress—start with zero and go up to design vibration stress in four or five steps.
- ❑ Repeat with other single stresses, such as humidity, voltage, transients, and shock.
- ❑ If there are no failures, continue the single stress cycles for a few hours. If there are still no failures, stage 1 is complete and we can proceed to stage 2.
- ❑ If there is even a single failure, validate the effectiveness of the correction using a B vs. C test (described in Chapter 11).

Stage 2: Single Stress Up to the Maximum Practical Over Stress Limit (MPOSL)

Purpose

Stage 2 is also a preliminary stage to determine the effect of overstress of each single stress used in stage 1. It need not be repeated for similar products, but only when a new product platform is launched or new environments encountered. The process is as follows:

- ❑ Continue stage 1 beyond the design limit for each stress on the same units of stage 1 (if repairable or, on new units, if not repairable) up to MPOSL. (It is not advisable to go all the way to destruct limit in each stress unless you must determine, empirically, what the destruct limit is likely to be for each stress.)
- ❑ If there are no failures in the overstress region, continue testing for a few cycles for a few hours.
- ❑ If there are still no failures, we can conclude one of the following:

  a. The stress type is inadequate.

  b. The rate of stress increase is too slow.

  c. The test has not been executed properly.

  d. The interaction failures between stresses have not been generated because of the absence of multiple environments.

  e. The design is robust for that stress.

- ❑ If there is only one failure per each failure mode, ignore the failure as a freak failure distribution or tail.
- ❑ If, however, there are two or more failures per each failure mode, analyze and correct each failure.
- ❑ Validate the effectiveness of the correction using a B vs. C test.

Stage 3: Prototype–Full MEOST to Maximum Practical Over Stress Limit (MPOSL)

Purpose

This is the most important of all MEOST stages—absolutely necessary before a design can be ready for production. When completed, it assures the designer of the best advance possible in reliability by forcing the weak links in design to be smoked out and then corrected. Stage 3 proceeds as follows:

- Select the four or five most important stresses that are likely to impinge simultaneously on the product in the field.
- Prepare a combined stress sequencing roadmap, along the lines of the examples in Figure 10-6 and Figure 10-7.
- Use the same units that have survived stages 1 and 2, if possible, and subject them to the combined stresses.
- Start at the design stress, then step-stress in four or five percentage increase intervals up to the MPOSL.
- The procedure is similar to stage 2, but with these multiple stresses.
- If there are no failures, continue the cycles of stress for at least twenty-four hours.
- If there are two or more failures per failure mode that are different from the most predominant failures in the field, then there are several possible reasons:

    a. The field data is old or unreliable, or product improvements have been made since the time the field data was gathered.

b. One or more stresses need to be added to the list.
c. The stress levels or the rate of stresses need to be increased.

- ❑ Perform another round of stage 3, with deliberately introduced "seeded defects" to confirm the effectiveness of MEOST in detecting them.
- ❑ Design improvements to correct the above failure mode(s) and validate the effectiveness of the improvement with a B vs. C test.

Sample Size

- ❑ Because there are only a few prototype samples that can be spared for any kind of testing, the sampling is 3 to 5 units.

Stage 4: Pilot Run

Purpose

This stage of MEOST ensures that design improvements/changes, tooling, suppliers, processes, and fixtures have not adversely affected design reliability. The steps are as follows:

- ❑ Run a stage 4 MEOST, using the same guidelines as stage 3, with new units from an engineering or production pilot run.
- ❑ A successful outcome means that the design is now ready for full production.

Sample Size

3 to 5 units

Stage 5: Mini-MEOST in Outgoing Production

Purpose

Stage 5 ensures that reliability integrity of the design is not degraded by manufacturing processes, workmanship, and supplier materials. The steps are as follows:

- ❑ Repeat stage 3, with two major exceptions:
  a. Reduce one or more of the four or five stresses used in stages 3 and 4.
  b. Reduce the overstress from the Maximum Practical Over Stress Level (MPOSL of stage 3 or 4) to the operational level (approximately one-third above the design stress).

Sample Size

- ❑ 3 to 5 units for production runs of 100 to 1,000 units per day; never use 100 percent sampling as in burn-in or HASS.
- ❑ 1 to 3 units for production runs of less than 100 units per day.

Frequency Sampling

- ❑ Once per week for production runs over 100 units per day.
- ❑ Once per month for production runs of less than 100 units per day.

Corrective Action

- ❑ If there are two or more failures per failure mode, production must be stopped until the effectiveness of the corrective action is verified with a B vs. C test.

Proof of Mini-MEOST Effectiveness

- ❑ After the first mini-MEOST, perform one round of stage 5 with "seeded defects" deliberately introduced.

Stage 6: First Round of MEOST on Field Returns

Purpose

Stages 1 to 5 are all performed virtually at time zero in their product life. Hence, plotting the failures or stresses on a Weibull plot will yield only one point on the graph and cannot be extrapolated to determine reliability over months or years of service. Stage 6's purpose, then, is to secure a second point on a Weibull plot after a period of exposure in the field—typically six months in service. The process works as follows:

- ❑ Make arrangements with a trusted, competent customer to keep track of field conditions, such as proper installation, preventive maintenance, and the scheduling of such maintenance, and record as much field history as possible.
- ❑ Request 5 to 10 good units to be returned from this customer in exchange for new units of the same product.
- ❑ Subject these units to a stage 3 MEOST, starting with design stress and continuing to MPOSL.
- ❑ Record the percentage of overstress when there are two (or more) failures. (Remember that design stress is 100 percent, while operational stress at 130 percent and MPOSL at 170 percent stress levels are typical.)
- ❑ On a Weibull plot, record the time to failure (zero time) on the X-axis and the percentage of overstress on the Y-axis from the previous stage 3 results.

- ❑ Similarly, record the time to failure (six months) on the X-axis and the percentage of overstress on the Y-axis from this stage 6.
- ❑ The end result is that we have at least two points on the Weibull chart.

Stage 7: Second Round of MEOST on Field Returns

Purpose

It takes a minimum of three points on a Weibull plot to draw a best-fit straight line connecting them and extrapolating the straight line (i.e., extending it in time) until it crosses the design stress level on the Y-axis. The corresponding point on the X-axis is the projected reliability of the product (in years). The purpose of stage 7 is to secure a third point on the Weibull after a second period of exposure in the field—typically one year in service—and to project a firm reliability figure for the product with which management can feel comfortable. Stage 7 proceeds as follows:

- ❑ Repeat the stage 6 MEOST regimen, only this time retrieve field units that have been in service for one year, instead of six months.
- ❑ On the Weibull plot, add a point depicting one year in service on the X-axis and the stress to failure of stage 7 on the Y-axis.
- ❑ With three points on the Weibull plot—at time zero, six months, and one year— draw a best-fit straight line and extrapolate it until it intersects the design stress horizontal line. Projected onto the X-axis, this intersection records the years to failure and hence depicts the reliability of the product. A simulated Weibull plot is shown in Figure 10-8.

Figure 10-8. Stress to failure over time.

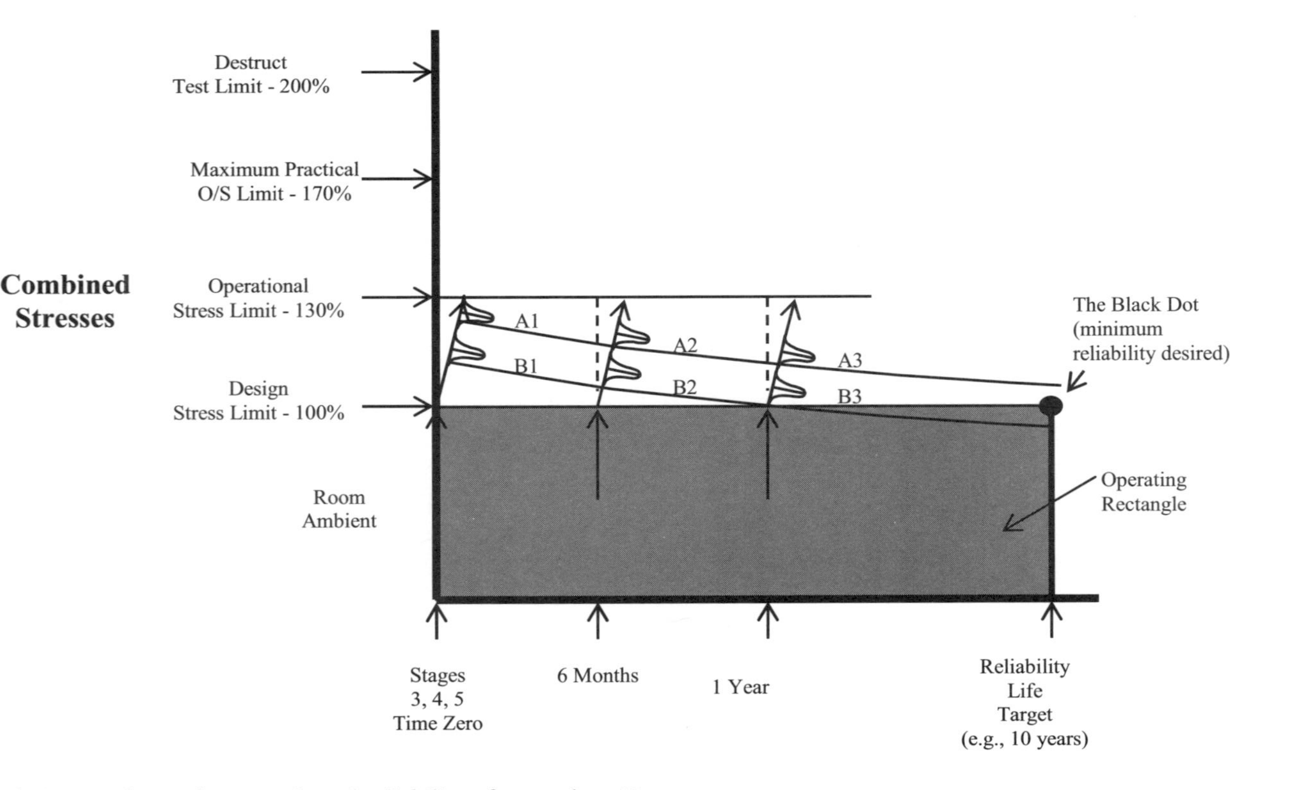

Conclusion: Product A has a projected reliability of more than 10 years.
Product B has a projected reliability of only one year.

- ❑ Continue stage 7 beyond one year, if desired, with products in service for two and even three years, to get a better confidence on the extrapolation.

Workshop Exercise

A truck manufacturer conducted a full-fledged MEOST study of its new air conditioner design. There were two principal failure modes:

1. Junction Block
2. Condenser

The results of the stress levels percentages for second failure on each failure mode were as follows:

		Stress Levels at 2nd Failure	
MEOST Stage	*Time*	*J-Block*	*Condenser*
3	0	170%	150%
6	6 Months	145%	141%
7	1 Year	125%	130%

Questions

1. Plot these figures on the Weibull chart in Figure 10-9, relating stress levels to failure and time.
2. Extrapolate and determine the reliability of each component of the A/C system, in terms of years (to reach 100 percent levels).
3. Does the J-Block subsystem meet the requirements of a three-year warranty? What is the expected reliability in years?
4. Does the condenser meet the requirement of a three-year warranty? What is its expected reliability in years?

Figure 10-9. Workshop Exercise: Weibull chart.

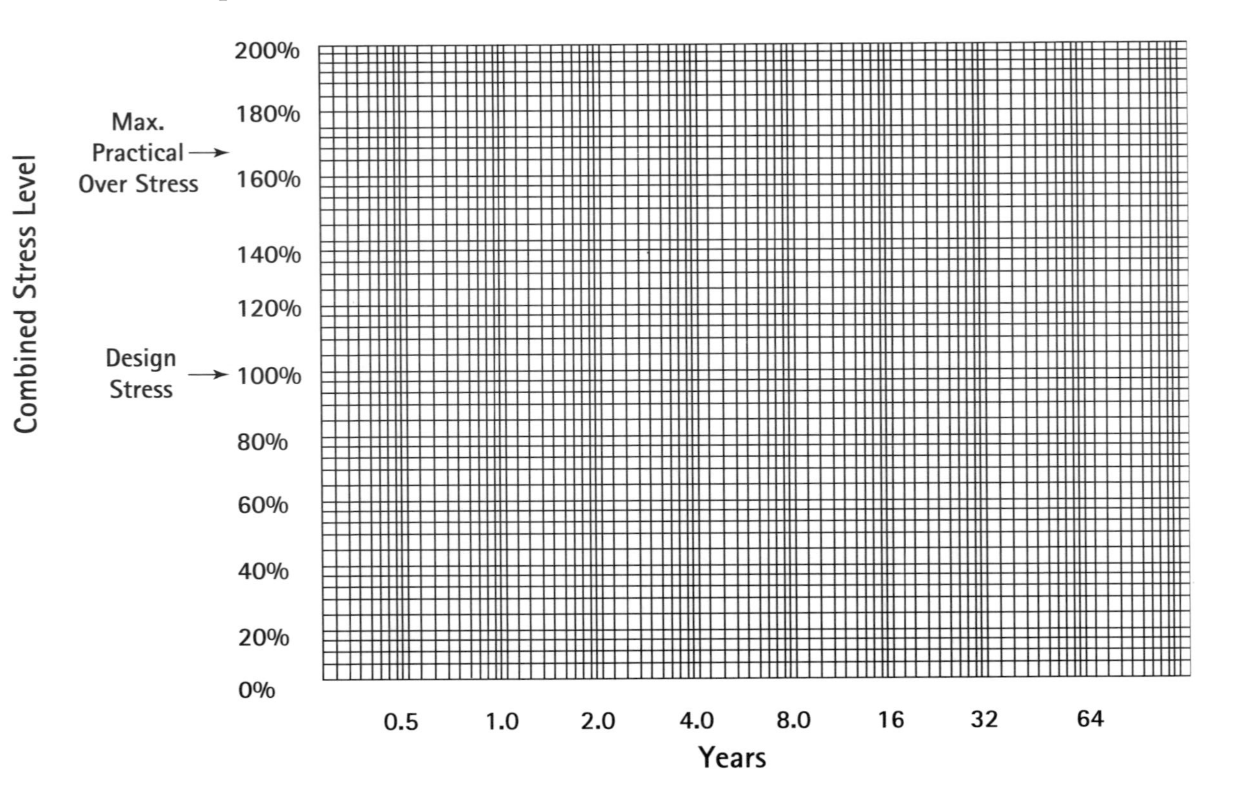

Answers

As shown in Figure 10-10:

1. The J-Block system will not meet the requirement of a three-year warranty. The Weibull extrapolation for a design stress level of 100 percent shows only a two-year life.
2. The condenser will easily meet the requirement of a three-year warranty. The Weibull extrapolation for a design stress level of 100 percent shows an eight-year life.

Stage 8: Cost Reduction

Purpose

Once the reliability of the product is improved with the previous seven stages of MEOST, company management has a firm number for reliability (which it always demands). But engineers and management may argue that promising reliability with MEOST can escalate product costs. Stage 8 provides the answer to these concerns. The motto is: Reliability first, through the seven stages of MEOST, then cost reduction at stage 8. There should be no attempt to reduce cost at the expense of reliability.

- ❑ Using a value engineering approach, proceed as follows:

    1. List all the high-cost items in the product and prioritize them in terms of highest cost and importance.

Figure 10-10. Workshop Exercise: Weibull chart, with data plotted and extrapolated.

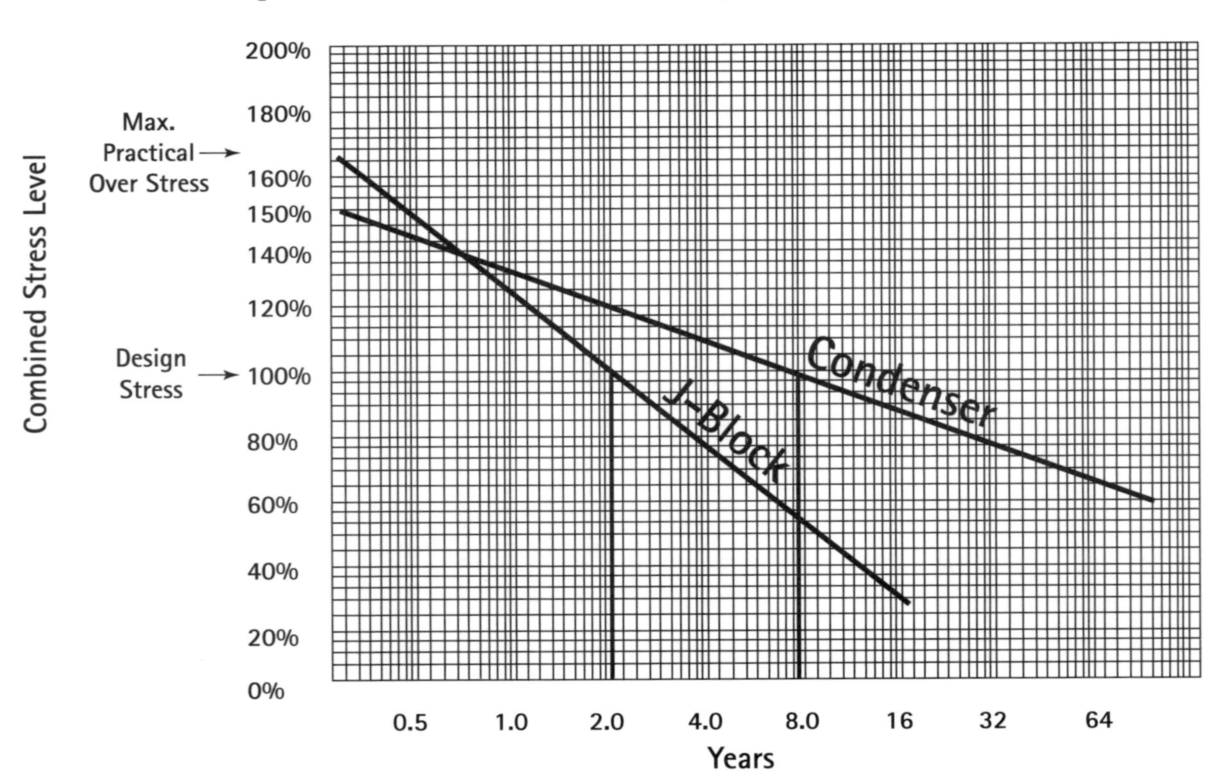

2. For the top three or four of the high-priority items, use a value engineering methodology to determine their function.
3. Using brainstorming and other related methods, determine what other part will provide the function at less cost. Estimate the potential savings.
4. Substitute the value-engineered part (or parts) for the more expensive original parts and run a functional test. Label this newer product as "B," for better, and the current product as "C," for current.
5. Run MEOST studies on both B and C products with three Bs and three Cs *in random order.*

- ❑ Note either a) the time to stress failure on all 6 units, or b) the stress to failure on all 6 units.
- ❑ Rank-order the results in terms of time to failure or stress to failure.
- ❑ If all three Bs outrank all three Cs (i.e., longer time or stress to failure), the improved lower-cost design is better than the current design with 95 percent confidence.
- ❑ If all three Bs are worse than all three Cs (i.e., less time or stress to failure), the new design should be rejected.
- ❑ If there is a mixed rank of Bs and Cs (i.e., some Bs are better than some Cs, and vice versa), then the new design is no worse than the current design and can be put into effect with the attendant cost-savings.

An example will illustrate the simplicity and power of the B vs. C test.

The drum of a washing machine was a high-cost system that the engineers wished to reduce in cost (C product). A lighter metal was tried (B product). MEOST stage 3 studies were conducted on three C products and three B products *in random order of testing* (in order to give all uncontrollable or "noise factors" an equal chance of entering or not entering the studies). The results of the six MEOST runs were as follows:

Scenario 1. Stress to Failure (Design Stress = 100%)

150% C

162% C

166% B

173% B

180% C

190% C

Conclusion: B is no better than C; nor is B worse than C. So the newer, lighter metal design would not degrade the product.

Scenario 2. Stress to Failure (Design Stress = 100%)

110% B

125% B

130% B

150% C

160% C

170% C

Conclusion: All three Cs outrank all three Bs. B is worse than C, with 95% confidence. The new design method should be rejected.

Scenario 3. Stress to Failure (Design Stress = 100%)

130% C

145% C

150% C

170% B

175% B

180% B

Conclusion: All three Bs outrank all three Cs. So the new design is not only less expensive, it's more reliable than the current design.

Conclusion

The seven stages of MEOST, along with stage 8 to explore cost-reduction potentials of the design, constitute the best, quickest, cheapest, and most powerful approaches to reliability enhancement known. It is also a recipe for getting a truly mature product into the marketplace way, way ahead of the competition.

CHAPTER 11

The Amazing Versatility of MEOST—New Applications and Challenges

"In terms of versatility, MEOST is the Michael Jordan in the court of reliability."

—NICK VON BAILLOU

The Economic Benefits of MEOST

Chapter 2 enumerated the benefits of reliability as a key corporate strategy. Together with quality, cycle time (especially design cycle time), and supply chain management, reliability constitutes a four-engine thrust to propel a company into a world-class stratosphere.

The pivotal role of MEOST, as the workhorse of reliability, has been described in the preceding chapters. The main points are once again outlined in this concluding chapter for recapitulation and emphasis.

1. *Contribution of MEOST to Profitability.* Corporate profits can soar with the application of MEOST. This has several dimensions:

   - ❑ Reducing warranty costs by factors of 10:1 and 100:1 (there is a multiplier effect on warranty costs, of 30:1 or up to 100:1, when the costs of customer defection, lawsuits, etc., are considered).
   - ❑ Reducing design test cost (in terms of the number of product samples and test facilities, and manpower requirements).
   - ❑ Reducing design cycle time (in terms of testing from months to hours).
   - ❑ Reducing manufacturing cycle time (through use of mini-MEOST) and the elimination of 100 percent testing, such as burn-in and HASS.
   - ❑ Increasing resale value in several important products (e.g., for automobiles, loss of resale value is $5,000; for trucks, it is nearly $10,000).
   - ❑ Reducing fruitless hand-holding trips to customers and dealers.
   - ❑ Reducing retrofit kits for the field.
   - ❑ Reducing product recalls (e.g., in the world of the Big Three automakers, more cars are recalled each year than are sold each year).

2. *Contribution of MEOST to Cost of Poor Quality Reduction.* Cost of poor quality (COPQ) consists of external failures, internal failures, appraisals, and prevention. These elements, added up, account for 10 percent to 20 percent of a company's sales—all of which is money down the rat-hole!

MEOST dramatically reduces:

- ❑ The cost of external failure (e.g., costs of warranty, buybacks, field trips, dealer giveaways, service giveaways, parts giveaways, and customer defections are enormous)
- ❑ Scrap and rework costs
- ❑ Lengthy and unproductive design validation costs

3. *Contribution of MEOST to Customer Retention.* It is well known that MEOST, by slashing field failure rates, increases customer loyalty, customer retention, and customer longevity. Furthermore, by improving reliability significantly, MEOST reduces product downtime. That's important to customers because downtime can have serious financial repercussions, ranging from the loss of a truckload of perishable food to millions of dollars of losses for electrical power companies for a few minutes of power outage in a whole city. Most shocking of all the reliability failures was the massive failure of the electrical power grid in the northeast United States and two Canadian provinces in 2003 that added up to billions of dollars!
4. *Contribution of MEOST in Design.* Speed in design is at least four times as important as design cost. MEOST significantly reduces the design validation tests from months to hours; reduces the total number of units required for such validation by factors of more than 100:1; reduces the headcount in engineering by at least 4:1; and reduces test cost by well over 10:1.
5. *Contribution of MEOST to Competitive Advantage at the Marketplace.* In the final analysis, a company that launches a design into the marketplace first is a winner. A company that is both first and has the most reliable product at the

same time is a triple winner (the third win is the synergy between marketability and reliability). With this awesome combination, competition can never catch up.

The Many Uses of MEOST—Its Commandeering of Many Disciplines

Aside from the powerful economic benefits that it brings to any company experiencing financial doldrums, MEOST also contributes to strengthening several disciplines and business practices in industry.

1. Problem Solving: Allied with Design of Experiments

It is a fact that 99 percent of industry does not know how to solve chronic quality problems. (The emphasis is on the word *chronic,* where problems fester for months on end.) The best way—no, the only way—to solve such problems is with the Shainin/Bhote Design of Experiments (DOE),[3] not Classical or Taguchi DOE.

MEOST, acting as the Green Y (i.e., the output or response) of DOE, can be used with Shainin/Bhote techniques, such as B vs. C, Variables Search, Paired Comparisons, and Scatter Plots—all DOE tools—to solve problems, especially if they occur only in the field and not in production. The most frequent MEOST Green Ys are 1) time to failure and 2) stress to failure.

2. Cost Reduction: Allied with Value Engineering

Value engineering—a powerful discipline, but one that's almost totally forgotten in most of the world except Japan—is a far more effective technique than traditional cost reduction. But in its practice, there is a tendency to cut costs at the expense of quality and reliability.

Used with MEOST, however (as seen in MEOST stage 8, described in Chapter 10), a natural constraint is placed on value engineering excesses by disallowing any cost reductions that adversely affect reliability. Unfortunately, very few MEOST practitioners go beyond stage 5 of MEOST, so they miss out on the economic benefits of cost-reduction stage 8.

3. Evaluating Improvements—How to Verify?

President Reagan once said about the Soviets: "Trust, but verify." Customers may trust a company's product improvements; manufacturing may trust engineering changes and improvements; engineering and quality assurance may trust a supplier's improvements. But how to verify these improvements? How can you be sure that they are both effective and permanent?

In facing a chronic reliability problem, the design team may assure an irate customer that its tests indicate an appreciable improvement. But six months to one year later, the improvement has vanished. Does the customer wait another six months to validate a second round of improvements? And what if the dreary cycles of improvements, launched every six months, are no better than the previous company promise?

In MEOST, we have a tool that not only predicts reliability in hours instead of months, but also validates a design change or a supplier change in hours instead of months. The test is a simple B vs. C (B for a better product, C for the current product). Three Bs and three Cs are subjected to a MEOST comparison, where all six units are stressed to failure (in random order) in a matter of hours. If all the three Bs show a higher time to failure or a higher stress to failure than all three Cs in the ranking of the six units, we obtain a 95 percent confidence in the improvement. If the ranking is mixed between the Bs and Cs, or all three Cs outrank all three Bs, there is no improvement.[3] The B vs. C technique is especially useful when the response (i.e., the Green Y) is an attribute (go, no-go, pass-fail), a yield, or if the defect rate is very low—say, below 100 parts per million (PPM).

This B vs. C Design of Experiments, coupled with MEOST, is an amazingly simple yet powerful technique for verifying improvements. Yet rather than implement it, the industry muddles along, going from failed promise to failed promise until, eventually, the customer votes with his feet to go over to a competitor.

4. Tackling the "No Trouble Found" Phenomenon with MEOST

An age-old problem for industry is the return of a product from a customer labeled defective, for which the company can find no fault. These "no trouble found" (NTF) or "checks OK" or "no apparent defect" (NAD) occurrences are the bane of almost all companies, especially those in the electronics business. Anywhere from 20 percent to 30 percent of all customer complaints are dismissed as NTFs.

The knee-jerk reaction of companies to NTFs is to:

- ❑ Ignore these NTFs and not include them as a part of warranty, or as a part of a company's responsibility.
- ❑ Blame the customer for not being familiar with the product's workings.
- ❑ Blame the service writer or service station for improper or inadequate diagnostics of the customer's complaint.

- ❑ Chalk it up to an intermittent problem that somehow got cleared.
- ❑ Accuse the service station of sending in false claims.

The first reform the company must make in its approach to NTF is to recognize that the customer complaint is indeed genuine; that an inadequate diagnosis has been made; and that an intermittent problem must be pursued to convert it from a "latent" defect to a "patent" defect, from a "soft" failure to a "hard" failure.

This is best done by stressing an NTF complaint with a MEOST study and forcing an intermittent failure into a permanent failure. Not all intermittent failures can be converted into such hard failures, but our experience indicates that at least 50 percent of NTFs can be smoked out as genuine failures with MEOST.

5. Converting Breakdown Maintenance into Preventive Maintenance with MEOST

Industry is notorious for ignoring action on a process or machine until there is a breakdown. The old motto "Don't fix it until it's broke" is still the norm. As a result, the ratio of breakdown maintenance to preventive maintenance is still 80:20 instead of a desired ratio of 20:80. Furthermore, the overall equipment effectiveness (OEE) of American industry is still in the 20–50 percent range, as compared to the best Japanese companies that have skyrocketed to the 85–95 percent range.

MEOST can pinpoint process variables that need to be changed at stated intervals before a breakdown occurs. The overstress testing can identify which process variables are likely to fail and in what order, so that a preventive maintenance can be established.

6. More Robust Designs with MEOST

Taguchi DOE attempts to make designs more robust with an orthogonal array, which combines the inner array factors (control factors) with outer array factors (uncontrollable noise factors). Besides the huge quantity of testing—say, six control factors multiplied by eight noise factors, which then require forty-eight experiments—and the complexity, the fraction factorial approach, which is the underlying methodology, is a flawed design that confounds (and contaminates) main effects with interaction effects.

Instead, MEOST can be used as a Green Y (response) to trade off uncontrollable noise factors against control factors, using a Variables Search experiment even as early as the prototype stage of design.

7. Out-of-Warranty Visibility and Reduction with MEOST

The period beyond specified warranties is a black hole for most companies. They have little knowledge and even less responsibility for failures that extend from the end of warranty up to the lifetime of the product. Yet these failures are important to customers. Manufacturers try to placate customers with extended multiyear contracts for which more money is extracted from the hapless customer.

MEOST brings back good units from the field six months, one year, and two years later for further stress tests (stages 6 and 7). As a result, not only does a company have visibility on potential failures five, ten, and fifteen years down the road, but it also has the means to correct such failures and reduce out-of-warranty costs for the customer.

8. Reducing Inventories of Spare Parts with MEOST

Many companies run a thriving business on spare parts. One company had a profit loss of more than 10 percent, but its spare parts business showed a very handsome 32 percent return on investment. That is the worst type of strategic thinking.

One of the features of MEOST—some may call it a weakness—is its ability to vastly improve reliability and, thus, shrink the need for and traffic in spare parts. Typically, spare parts management suffers from two diametrically opposed conditions. It is faced with spare part shortages on high-failure-rate items while it has excess inventory on other parts.

By pinpointing high-failure items early in the design and correcting them, MEOST reduces the risk of high-failure rates on these items and the need to stock them. In addition, by identifying the strong parts that do not fail in overstress tests, MEOST can reduce the inventories of such parts to barebones minimum.

9. Increasing Supplier Profitability with MEOST

MEOST as a profit-generating technique for a company has already been discussed. But its profit potential for key suppliers is equally rosy. Field success requires that a company make reliability a quantifiable and measurable specification for key parts for key suppliers. Strengthening this requirement with financial incentives for suppliers that meet the specification, and financial penalties for those that don't, is a practice whose time has come.

MEOST is the best way for a supplier to guarantee that his parts can maximize such financial incentives and minimize, or even eliminate, the risk of financial penalties. The supplier, in turn, can impose similar financial incentives and penalties on his suppliers, thus strengthening the entire supply chain with the bonanza of MEOST.

10. Reducing the Risk of Product Liability and Lawsuits with MEOST

Product recalls, product liability problems, and their attendant lawsuits are, unfortunately, becoming a way of life in the litigious culture that is America. Ours is less a nation of laws than it is a nation of lawyers who have their sticky fingers in the pockets of one of our two major political parties, so tort reform is not likely to come into being.

Under such trying circumstances, a company has no alternative but to guard against product liability exposure, no matter how frivolous a case against it might be. Chapter 9 dealt with product liability prevention and what a company can do to protect itself from being dragged into court and, possibly, jail. Here again, MEOST can be a powerful ally arrayed against predator lawyers.

In MEOST experimentation, seeded problems, such as potential customer misuse, critical part failures, and other concerns listed in failure mode and effects analysis (FMEA) studies, can be deliberately introduced to see the consequences of such problems. The design can then be strengthened to overcome them. Furthermore, the very act of simulating all such possibilities at the design stage can be impressive court testimony for the defense in a lawsuit.

Conclusion

It is our hope that this book will convince engineering managers and the engineering community to jettison their antediluvian methods of predicting, estimating, and demonstrating reliability as well as validating their designs. Instead, we have shown them new ways to discover field failures in hours instead of weeks and months of testing and, in the process, beat their competition to the punch.

This book will, we believe, enlarge the horizons of quality professionals to the new world of reliability, with MEOST as its centerpiece, and enhance their primary mission, which is to act as coaches to their engineering, manufacturing, and supplier constituencies instead of performing sterile policemen roles.

Finally, it is our hope that this book will spark the interest of the top management of companies, especially those that are mired in mediocre and shrinking profits, to find a new path to a bottom line that's three to four times larger than their current figures.

Appendix: Laboratory Equipment/Chambers for MEOST

The Management Dilemma

One of the major obstacles in launching MEOST in a company is convincing management that sizable capital investment—ranging from $50,000 upwards to $500,000—will be needed to equip its own laboratory. This is not the first battle that proponents face in loosening management purse strings for any kind of capital investment. Nor will it be the last. But, in the judgment of the authors, there are few capital expenditures where the savings in warranty, retrofits, product recalls, resale value, and product liability lawsuits are at least an order magnitude—perhaps two orders of magnitude—over the investment.

Yet, how to convince a nervous management of the wisdom of such an investment? Most managers are afraid that overstress is overkill. Yet, there are a couple of ways to reassure them.

One method would be to benchmark those companies that have made these MEOST-type investments and learn from their success stories. Today, there are several companies that, if they have not gone all the way to MEOST, have at least gone halfway by adopting HALT and HASS techniques. These end-user companies can vouch for the general effectiveness of overstress tests. Their numbers are constantly on the rise.

A second method—one that the authors strongly suggest—is to select a company's product that has experienced a high failure rate in the field, with known failure modes, and select an independent laboratory that has MEOST-type equipment for a MEOST trial in the laboratory. There are a few such facilities, including the manufacturers of these overstress chambers, such as Thermotron Industries, QualMark Corp., and Entela, Inc. (Care must be taken, however, to avoid those laboratories that do military testing with AGREE chambers. They may claim, and do, that the stringent military requirements should be "good enough" for any MEOST testing. They are shocked when we tell them that military test requirements are antiquated and inadequate for MEOST.) The proof of the pudding is for the laboratory to reproduce the same failure modes, which took months and years in the field, in a matter of eight to twenty-four hours!

Once there is proven success with such independent test laboratories, capital investment in MEOST should be a no-brainer. (At Motorola's Automotive and Industrial Electronics Division, our first MEOST chamber, costing $250,000, took us one full year to secure management approval. A year later, the second chamber, costing the same amount, took only one week!)

Traditional Equipment for Reliability Just Will Not Do

Traditional test equipment for design validation includes very slow rates of temperature change, generally 2° C per minute, with mechanical compressor systems and electrodynamic vibration, as well as chambers that are incapable of combining stresses. The results are poor reliability prediction and demonstration, and even poorer field reliability.

Major Elements of MEOST Chambers

MEOST chambers combine, at the very least, temperature cycling and vibration. To these stresses are usually added humidity, voltage transients, power cycling (on/off), and frequency cycling. It is not practical to combine more than four or five stresses. As a general rule, special environments that are independent of other environments should be tested separately, one environment at a time.

Thermal Cycling

MEOST chambers should be capable of meeting these five requirements for temperature cycling:

1. *Temperature Range:* –65° C to +180° C.
2. *Rate of Temperature Change:* A minimum of 25° C per minute and, preferably, 60° C per minute.
3. *Dwell Time:* As short as possible to allow key product areas to reach the same temperature as chamber ambient.
4. *Turbulence:* Air turbulence, rather than an even airflow rate, is necessary to make sure that air pockets causing hot or cold spots in the product are eliminated (200 cubic feet/minute is recommended).
5. *Liquid Nitrogen (not mechanical) Compressor:* The technical and economic advantages of liquid nitrogen (LN_2) as compared with a mechanical compressor system are

shown in the following table. However, since liquid nitrogen can, under certain circumstances, deplete the oxygen in the room, the addition of an oxygen depletion monitor is necessary for safety reasons.

Features	*Liquid Nitrogen* LN_2	*Mechanical Compressors*
A. *Technical*		
Reliability	Excellent	Poor
Size	1X	2X
Weight	1X	2X
Noise	Lower (60 DBA)	Higher (80 DBA)
Pollution	None	Some
Safety	Danger of O_2 Depletion*	Freon a Hazard
B. *Thermal*		
Range	–100° C to +200° C	–40° C to +170° C
Ramp Rate	60° C/Minute	30° C/Minute
C. *Cost*		
Purchase	1X**	2X**
Operational	1X	6X
Maintenance	1X	200X

*Requires an oxygen depletion monitor.
**X is a multiplier.

Vibration

MEOST chambers should be capable of meeting these ten characteristics for vibration (simultaneously with the thermal cycle requirements):

1. Vibration system: Random, repetitive pneumatic shock.
2. Degrees of freedom: Six (three orthogonal axes and three rotational vectors, using each of three axes simultaneously).
3. Frequency range: 2 to 5,000 Hz.
4. Acceleration: 2 grms to 40 grms.

5. Vibration levels should be higher at edges of product than in the middle.
6. High frequencies create more failure modes in low-mass products (e.g., surface-mounted components).
7. Low frequencies create more failure modes in high-mass products.
8. Products and fixtures do attenuate the original vibration energy at the table.
9. The frequency bandwidths should go up to the full 5 kHz level. This is especially important to discover surface-mounted devices, internal bonds, and solder joints.
10. For large mass components, with large displacements, it may be necessary to follow up MEOST with more traditional vibration systems, such as the electromechanical type.

General Considerations

- ❑ Double doors are preferable for MEOST chambers for easy access to the product by operators and servicers.
- ❑ Ducting the air across the product, with turbulence, greatly improves the airflow and reduces test times.
- ❑ The chambers should have multiple air discharge ports with balanced airflow. Attaching simple flexible ducting ensures efficient testing.
- ❑ It is preferable to remove the outer cover of the product to ensure that the airflow within the product is continuous and unrestricted.
- ❑ It is advisable to place a thermocouple in the duct's air path to prevent overheating.

- ❑ Chambers with low noise levels at full vibration levels will accommodate conversation among operators.
- ❑ Quick serviceability must be assured, especially for the vibration equipment.
- ❑ Interconnections between the product and the test equipment should use Teflon cables rather than the usual PVC jacket cable. Temperatures over 90° C in PVC can produce toxic and even fatal gas (chlorine).
- ❑ The equipment should feature lifting mechanisms to reduce heavy weight lifting by operators and promote safety.
- ❑ If electrostatic discharge (ESD) is important for the product, it is recommended that a conductive flooring system that meets ESD requirements be installed for the chamber.

References

1. Keki R. Bhote, *The Power of Ultimate Six Sigma* (New York: AMACOM Books, 2003).
2. Keki R. Bhote, *Beyond Customer Satisfaction to Customer Loyalty* (New York: AMACOM Books, 1995).
3. Keki R. Bhote and Adi K. Bhote, *World Class Quality,* Second Edition (New York: AMACOM Books, 2000).
4. Ibid.
5. Seiichi Nakajima, *TPM Development Program* (Cambridge, MA: Productivity Press, 1989).
6. Nachi Fujikoshi, *Training for TPM* (Cambridge, MA: Productivity Press, 1990).
7. Shigeo Shingo, *Zero Quality Control, Source Inspection and Poka-Yoke* (Cambridge, MA: Productivity Press, 1988).
8. Nikkan Kogyo Shimbun, *Poka-Yoke: Improving Quality by Preventing Defects* (Cambridge, MA: Productivity Press, 1988).
9. Keki R. Bhote, *Strategic Supply Management* (New York: AMACOM Books, 1989).
10. Keki R. Bhote, *The Ultimate Six Sigma* (New York: AMACOM Books, 2002).

11. Ibid.
12. Geary Rummler and Alan Brache, *Improving Performance—How to Manage the White Spaces on the Organization Chart* (San Francisco: Jossey-Bass, 1990).
13. Marketing Concepts, Inc. "Survey of 5,873 Consumers."
14. Kam L. Wong, "The Bathtub Curve Does Not Hold Water Anymore," Quality and Reliability Engineering Symposium (1988).
15. Kam L. Wong, "What Is Wrong with the Existing Reliability Prediction Methods?" *Quality and Reliability Engineering International* (1991).
16. Thermotron Industries, "Fundamentals of Accelerated Stress Testing" (1998).
17. CALCE Electronic Packaging Research Center, "Environmental Stresses and Effects" (University of Maryland, 1995).
18. George Kujawski and Edward Rypka, "Sense and Nonsense in Environmental Stress Screening" (San Jose, CA: ESSEH, 1991).
19. Harry W. McLean, *HALT, HASS, and HASA Explained* (Milwaukee, WI: ASQ Quality Press, 2000).
20. Bob King, *Better Designs in Half the Time* (Salem, NH: Goal/QPC, 1987).
21. Geoffrey Boothroyd and Peter Dewhurst, "Product Design for Manufacturing Assembly," 1988 International Forum on DFMA (Wakefield, RI: Boothroyd Dewhurst Inc., 1988).
22. Nikkan Kogyo Shimbun, *Poka-Yoke: Improving Product Quality by Preventing Defects* (Cambridge, MA: Productivity Press, 1988).
23. Sam DeMarco, *Value Model for Accelerated Testing* (Southfield, MI: Society of Automotive Engineers, May 7, 2002).

Index